OVERTRAINING ATHLETES

Personal Journeys in Sport

Sean O. Richardson, PhD

Mark B. Andersen, PhD

Tony Morris, PhD

Victoria University, Melbourne, Australia
Centre for Ageing, Rehabilitation, Exercise, and Sport
and The School of Human Movement, Recreation, and Performance

Human Kinetics

Library of Congress Cataloging-in-Publication Data

Richardson, Sean O., 1972-
  Overtraining athletes : personal journeys in sport / Sean O. Richardson, Mark B. Andersen, and Tony Morris.
    p. cm.
  Includes bibliographical references and index.
  ISBN-13: 978-0-7360-6787-4 (soft cover : alk. paper)
  ISBN-10: 0-7360-6787-6 (soft cover : alk. paper) 1. Athletes--Training of. 2. Athletes--Psychology. 3. Physical education and training. 4. Sports--Psychological aspects. 5. Sports--Physiological aspects. 6. Sports medicine. I. Andersen, Mark B., 1951- II. Morris, Tony, 1950- III. Title.
  GV711.5.R53 2008
  613.7'11--dc22

                                        2007048251

ISBN-10: 0-7360-6787-6
ISBN-13: 978-0-7360-6787-4

Acquisitions Editor: Michael S. Bahrke, PhD
Developmental Editor: Kevin Matz
Assistant Editor: Laura Koritz
Copyeditor: Tom Tiller
Proofreader: Anne Rogers
Indexer: Craig Brown
Permission Manager: Carly Breeding
Graphic Designer: Fred Starbird
Graphic Artist: Kim McFarland
Cover Designer: Robert Reuther
Photographer (cover): Tim Tadder/Corbis
Photographer (interior): © Human Kinetics, unless otherwise noted
Photo Asset Manager: Laura Fitch
Photo Office Assistant: Jason Allen
Art Manager: Kelly Hendren
Associate Art Manager: Alan L. Wilborn
Illustrator: Alan L. Wilborn
Printer: Versa Press

Printed in the United States of America   10 9 8 7 6 5 4 3 2 1

Human Kinetics
Web site: www.HumanKinetics.com

United States: Human Kinetics
P.O. Box 5076
Champaign, IL 61825-5076
800-747-4457
e-mail: humank@hkusa.com

Canada: Human Kinetics
475 Devonshire Road Unit 100
Windsor, ON N8Y 2L5
800-465-7301 (in Canada only)
e-mail: info@hkcanada.com

Europe: Human Kinetics
107 Bradford Road
Stanningley
Leeds LS28 6AT, United Kingdom
+44 (0) 113 255 5665
e-mail: hk@hkeurope.com

Australia: Human Kinetics
57A Price Avenue
Lower Mitcham, South Australia 5062
08 8372 0999
e-mail: info@hkaustralia.com

New Zealand: Human Kinetics
Division of Sports Distributors NZ Ltd.
P.O. Box 300 226 Albany
North Shore City
Auckland
0064 9 448 1207
e-mail: info@humankinetics.co.nz

For my Mom and Dad, Hilde and John, who have supported and loved me through all of the ups and downs of my athletic and academic endeavours, who have been there for me when I was knocked down by my own overtraining, and who have given me confidence to accomplish anything and everything in life.

Sean O. Richardson, PhD

For all my athlete-clients who trusted me with their life stories and taught me more about overtraining, desire, and human frailty than any book ever could.

Mark B. Andersen, PhD

For my family, Felicity, Rachel, and Adam who have put up with my own extreme overtraining in sport psychology for way too long.

Tony Morris, PhD

Contents

PART III What Can We Learn From Athletes?

CHAPTER 7 The Pathogenic World of Professional Sport: Steve's Tale

PART IV Past Models and Current Conceptions · · · · · · · · · 167

Acknowledgments

Thank you to the athletes, coaches, and sport scientists who participated in the interviews, which formed the basis of both my doctoral dissertation and this book. Sharing the intimate details of your lives, your painful experiences with overtraining and injury, took courage; your legacy will stand to inspire future generations of athletes and coaches to find balance somewhere amidst their relentless pursuits of excellence. Thank you to the three examiners of my dissertation, Michael Kellmann, Andrew Sparkes, and Peter Hassmén, for kindly dedicating your time to reviewing, and to providing valuable feedback on, the doctoral dissertation. Thank you to Tony Morris and Mark Andersen for your guidance, both academic and personal, over the years it has taken to put my doctoral work, and this book, together. Tony, as my principal doctoral advisor, you have supported me through difficult times, challenged me to think expansively, kept me focused on the task, and allowed me the space to nurture and express my own ideas. Mark, as my doctoral co-advisor, and my practicum supervisor, you have driven me to achieve the highest standards possible in my writing and practice, opened my mind to new ways of thinking, and infuriated me along the way, but made me feel loved at the same time. Had it not been for you and your dogged determination to publish our work, which I did not want to look at ever again after finishing my doctorate, this book would not have become a reality. Thank you to the staff at Human Kinetics for supporting our project. If this book can help just one athlete make a difference in his or her life, then we have succeeded. Finally, thank you to my family, John and Hilde (parents), Sasha and Daniel (siblings), for always loving and supporting me in all of my life's endeavors.

Sean O. Richardson, PhD

I am a manic-depressive writer and editor. I generally have two speeds: full tilt and dead stop. My writing and editing psychopathology means that the people I work with either get bombarded with massive amounts of material to work on and are subjected to my impatience and nagging in the manic phase, or they want to start me on Zoloft to pull me out of my author's torpor and despair. It's the way I am built, and years of psychotherapy haven't helped much in this department. So I would like to thank those people who have had to put up with me while working on this book. First, I would like to thank Rainer Martens, who sat down with me at an Association for Applied Sport Psychology (AASP) meeting and asked what I thought needed to be published in the field. This book is a direct result of that conversation. Myles Schrag and Kevin Matz of Human Kinetics have been major anchors in the to-ing and fro-ing that accompanies creative and scientific endeavors. My co-authors, Tony and Sean, have been unbelievably sanguine and steady colleagues throughout this whole process. This book is Sean's baby; Tony and I were only his midwives. Thanks to Dr. Harriet Speed, or as we call her in Swedish, Fröken Fart (translation: Miss Speed). She is a constant source of both gut-busting hilarity and sage counsel. Sincere thanks to Trisha Leahy and David Martin who, for over a decade, have been great friends, colleagues, and storytellers. And finally, thanks to my partner, Jeffrey, for too many reasons to list here.

Mark B. Andersen, PhD

First, I wish to thank Sean, whose quest to understand his own story in elite sport introduced me to the fascinating, yet frightening, field of overtraining. To Mark, my appreciation for interpreting for me the dark side of elite athlete behavior that Sean explored, and for his huge, if manic, effort in bringing this book to fruition. Of course, I am very grateful to the athletes, coaches, and sport scientists who spoke so candidly to Sean, and later to Mark, and without whose input there would be no stories to tell. My thanks also go to Rainer Martens for his vision in supporting the concept of a book on this topic and to Myles and Kevin at Human Kinetics for their care and support during the editorial process.

Tony Morris, PhD

Introduction

This tale begins with a journey, as perhaps most do. It is a journey aimed at discovering what drives athletes to go over the edge in terms of injury, illness, and fatigue. It is also a bit of a personal journey for the first author, Sean. He was one of those athletes who went too hard, did not get enough recovery, and ended up injured. He loves competitive sport, and he loves winning, but he has started to see that the cultures and subcultures of competitive sport and winning don't always leave athletes happy and healthy, especially when they push their bodies and minds too far out of balance. Along his journey, Sean went through an evolution from competing as an enthusiastic young athlete, never having experienced debilitating injuries, and not having much sport science knowledge, to returning to competition following major injury, equipped with a solid sport science knowledge base, a master's degree, and most of a PhD on overtraining and injury. Sean believed he had developed the tools, through his experience and education, to balance his training, prevent injury, and avoid the potentially debilitating outcomes of overtraining. Nonetheless, in his final attempt to make the Canadian Olympic Rowing Team, Sean found himself off balance, pressured to push through pain and niggles by a system of sport in which athletes experience a surfeit of overt and covert forces to keep training in the face of pain, illness, and fatigue. Together with his coauthors of this book, Tony and Mark, Sean set out to understand why he and other athletes drive themselves to engage in overtraining behaviors and end up experiencing the associated negative outcomes. The motivation to comprehend the potentially complex interactions of variables that lead athletes to push too hard, neglect proper recovery, or shut out the consequences of injury, illness, and fatigue became the starting point for this project, and the foundation for this book.

Reflecting on Sean's experiences, and observing other sports, it appears that athletes often push themselves to the limits of their physical capacities, prompted by the belief that large volumes of training form the pathway to success. Such heavy training, however, can lead to the undesirable outcomes of decreased performance levels, illness, and injury.

Athletes and coaches commonly respond to such drops in performance, or setbacks from injury or illness, by increasing the training load still further. In many sports, athletes similar to Sean experience substantial direct and indirect pressures to perform from coaches, parents, administrators, and themselves, and often will do "whatever it takes" to win. The omnipresent sport culture, which envelopes coaches and athletes alike, drives them to embrace this more-is-better philosophy in training for peak performance, and the risk for an athlete to descend into a state of physiological exhaustion, illness, or injury escalates with mounting pressures and increasing workloads.

PERSPECTIVES ON TELLING TALES

Sean's life in sport as a competitor, a coach, an observer, and a sport psychologist—along with his own experiences with the phenomena of overtraining (OT) and injury—positioned him as an active participant in the research process of this project and influenced the ways in which the knowledge he gained will be represented in this book. Furthermore, Tony's and Mark's experiences as experts, both in working with and studying athletes for many years, have helped shape their contributions in guiding Sean's project, interpreting the stories of experts and athletes, and writing this book. All three authors will be written into the text, having their voices and experiences highlighted throughout the book at various times.

What we are interested in presenting within this book are the stories and insights that evolved from Sean's research, which included descriptive studies of athletes', coaches', and sport scientists' experiences with OT processes and outcomes in elite sport. Using two studies from Sean's research project—the first an interview study with sport experts, the second with athletes—and two follow-up interviews with experts from the original research, we hope to illustrate the variety of perspectives on and experiences with OT. After all the tales are told, we present a model of overtraining that encompasses the stories we have been privileged to hear, the past research, and the models currently in use. We also make recommendations for applications of the research and the model in managing overtraining, injury, and stress–recovery imbalances.

In telling the experts' and athletes' tales in this book, we have chosen a mix of presentation styles, both realistic and confessional. In particular, following the realist tradition, several sections of the chapters on the experts and athletes are dominated by quotations from the participants (Sparkes, 2002).

With the extensive use of participants' quotations, we hope to invite the reader to take part in their stories, and perhaps to identify with the athletes' experiences of overtraining or the experts' experiences of working with overtrained athletes. Nonetheless, with Sean's story of OT to tell, and with Mark's and Tony's expert (though rather academic, we readily confess) perspectives, we have also included our own voices—the confessional elements of telling the tales, especially with respect to retelling and interpreting the stories of the athletes, and offering personal reflections on the interview material.

Sean identified with Sparkes' comments on how confessional tales allow researchers to describe the evolution of their points of view through the research process:

> The . . . point of view is often represented in confessional tales as part of the character-building conversion tale in which the researcher, who had a view of how things might happen at the start of the study, comes to see things very differently as the study progresses. (p. 60)

Although the main thrust of the stories in this book comes from the participants' voices—from the experts' tales of working with OT athletes, and from the athletes' tales of OT and injury—we have filtered the tales of the experts and athletes through our own reflections, interpretations, and confessions. We hope that, by including our voices in telling the tales, the reader may understand more clearly the perspectives presented and find the journey through the complex phenomena of OT experiences more interesting and informative. We think it's a hell of a ride, and we hope you will enthusiastically join us.

EVOLUTION OF THIS PROJECT

In setting goals for the outcomes of this book and Sean's project, there was a shift in the focus of

the research from OT syndrome (OTS), a singular outcome, to OT processes and behaviors with multiple negative outcomes. We could most accurately describe the changes in perspective by saying our views have broadened and deepened. At the outset, it seemed we would be exploring only the risk factors for overtraining syndrome (a negative outcome associated with fatigue and underperformance) in competitive sport and that the project would be limited to sports typically associated with high volumes and intensities of physical training, such as swimming, cycling, and rowing. The focus has broadened, however, from trying to identify risk factors for this relatively infrequent outcome in elite sport, OTS, to looking at risk factors for an entire process encompassing OT behaviors and outcomes, including illness and injury, which are common experiences for most athletes. Our views have deepened in that we have come to see OT as a process with an elaborate set of behaviors and multiple outcomes, resulting from interactions among many aspects in athletes' lives. OT seems to be about more than just understanding how athletes handle a given training load; it encompasses seeing athletes as complex beings with personal histories, influenced by significant others, driven by internal conflicts, and pressured by sport cultural traditions.

What Sean discovered when he began talking with athletes is that, regardless of sport and intensity of training, most athletes could tell stories of times when they went too hard, exceeded their limits, failed to recover sufficiently, or became overstressed. They returned too early from injury; turned a blind eye to niggles that eventually got worse; acted desperately to achieve goals that might have been unrealistic; or made poor decisions about health, training, and injury, in general, when physically or psychologically vulnerable. The commonality among many of the stories told was that they were connected by underlying behavior patterns (e.g., that of doing too much, given the individual's capacity to cope). With a perspective focused on OT processes, one can see that it is possible for athletes to overtrain without high training volumes or intensities, and to risk negative outcomes beyond the narrow scope of OT syndrome, such as illness or injury, if they are compromised in other areas of their lives.

Understanding OT in elite sport is a complex issue. Many researchers, coaches, and sport scientists have commented on risk factors for OT and injury in athletes; some have offered illuminating anecdotes, and others have presented case-study findings illustrating experiences with OT. Kenttä and Hassmén (2002) have proposed a comprehensive model for understanding the overtraining and underrecovery process in terms of a stress–recovery balance. We developed a plan to interview elite athletes, coaches, sport doctors, psychologists, and exercise scientists. We thought we could gain insight into the personal and situational factors surrounding OT processes and outcomes by exploring multiple perspectives from athletes in the field who have experienced OT and from experts who have worked with athletes who have overtrained.

OUR GOALS FOR THIS BOOK

We endeavored to continue the trend, started by Kenttä and Hassmén (2002), of distinguishing causes and consequences of OT and extending the understanding of personal and situational risk factors. The field is moving from identifying OT syndrome once it has happened to anticipating OT behaviors and outcomes by looking at risk factors. With risk-factor research, it might be possible to answer a key question: Can one identify and take steps to change OT behaviors before they damage an athlete severely? This question is intimately connected to larger issues of happiness and quality of life experiences. Our aim with this book is to provide insight into OT experiences, so that athletes, and people working with them, can take steps to make the world of sport and competition a safer and more salubrious place to be. This book is for all people involved in sport: exercise physiologists, sport psychologists, sports medicine personnel (including rehabilitation specialists), coaches, athletes (and potentially their parents and significant others), and fitness professionals. They all have roles to play in the recognition and prevention of overtraining. We believe we have written a book that will be accessible to many of the people involved in athletes' lives.

We want the athletes we work with to do good things in their sports and in their lives. We are all part of the equation. We hope this book helps explain how all of us intersect in ways that help, hinder, and even damage athletes in their personal quests. In reading the tales told here and reflecting on them and on what they say about service and practice, we invite the readers to consider how best we can serve those athletes in our care—the girls and boys and women and men we invest in, hope for, and even (dare we say it?) love.

INTRODUCTION TO RESEARCH AND TERMINOLOGY IN OVERTRAINING

Introduction to Overtraining

Any Olympic year will provide several examples of the results of overtraining in elite sport. There are always Olympic stories of athletes who struggle with overuse injuries stemming, most likely, from precisely that: overdoing their training and damaging muscles, tendons, or ligaments. We also read about athletes who have fallen ill just before their major events. Why these illnesses occur, at exactly the wrong time, probably has to do with training too hard when already at maximum capacity, not getting enough recovery, depleting resources, and not managing the major stressors of going for Olympic gold. These factors affect immune system function, and many athletes may begin their Olympic experiences in immuno-compromised states.

Athletes need to train hard to reach peak performance, and one general principle known to all elite athletes and coaches training for major events is to *overload* the athlete. Athletes train hard and long, deplete resources, and do not recover adequately from one training session before starting the next one. In general, athletes become fatigued and performance does not improve (or actually drops off). This overload, however, is just the first step. What follows overload is the taper, wherein the athlete eases up considerably and takes plenty of time to recover, whereupon the body bounces back with a supercompensation response. This compensatory rebound may help athletes reach new heights in performance (see Steinacker & Lehmann, 2002).

If tapers do not work and athletes remain in fatigued and performance-suppressed states, then something has gone wrong with the overload

phase. In short, athletes can become *over-overloaded*, or, in another word, overtrained. This sequence of overloading and poor response to taper is, however, only one part of our overtraining story. In this book, we take a broad perspective on overtraining that covers physiology, psychosocial factors, personal histories, and influences of cultural context on overtraining outcomes. In the rest of this chapter, we will examine the research literature on overtraining and also focus on the use of terminology when examining overtraining phenomena.

PAST STUDIES AND LIMITATIONS IN OVERTRAINING RESEARCH

In the scientific literature, the negative processes and outcomes associated with excessive training load have been called *overtraining* (OT) and *overtraining syndrome* (OT syndrome), respectively. Parmenter (1923) identified OT as a thorny issue in competitive sports in the early 1920s, and Griffith (1926) referred to the negative outcome state associated with intensive training as *staleness*. Research on OT did not begin to accumulate, however, until the mid-1970s and early 1980s, when competitive athletes began to train at substantially greater volumes and higher intensities than their earlier compatriots. Bompa (1983) estimated 10% to 22% increases in yearly training hours for a variety of sports during the five years from 1975 to 1980. Between 1972 and 1995, American Olympic-level swimmers increased training loads from around 9,000 meters to 36,000 meters per day (Peterson, about whether OT may have positive or negative aftereffects; about whether it should be considered a process, an outcome, or both; about whether various aspects of OT are causes or consequences; and about the varied usage of terms in the fields associated with OT. We present definitions in the following sections, illustrating some of the subtle differences in various usages of OT terminology.

That overtraining occurs when things go awry in the training process is obvious, but we need to understand the context and processes of "normal" training in order to see where, how, and when athletes get off track. To begin with, it may be useful to look at a description of how the training process is viewed today. Steinacker and Lehmann (2002) outlined what training includes:

> Athletic training consists of repetitive phases of normal training, high-load training, overload training, overreaching, and recovery. During the training program, training load—defined by the intensity, duration, and frequency of exercise—varies and should gradually increase in response to the training-induced adaptation of various physical systems. This increase in training load is necessary to ensure further responses to a training program. Coaches often organize training in alternating cycles of increasing training load and enhancing regeneration. Such training cycles, which are relatively safe, allow the training load to reach a high, sustainable level for a short time. During the process (which is called supercompensation, or overreaching) the exhaustion and fatigue resulting from the high-load training phases elicit corresponding cellular stresses and consecutively [as a consequence of taper and reduced load] raise performance in the recovery phases as an adaptation to the training overload. (p. 103)

What seems evident from this description of the training process is that athletes are intentionally pushing their training hard to get optimum results. High levels of fatigue and physiological adaptations are to be expected, and the objective of the process is to achieve peak performance after a period of recovery (i.e., a taper). The results of periods of high-intensity training hinge on what happens during the taper. If athletes do not bounce back to previous levels of performance (or better) after a week or two of lower-intensity or lower-load training (tapering), then supercompensation did not take place. The athlete may have used so many

resources that recovery was impaired, and the overload period probably went on for too long.

Overtraining and Overtraining Syndrome

There is a problem with the word *overtraining*: Its definition varies depending on the author or researcher, so we need to sort out how others have defined this term and make clear how we will be using it in this book. The following section presents definitions of OT and OT syndrome that various researchers in the field have formulated. Steinacker and Lehmann (2002) offered the following:

> *Overtraining* is a long-lasting performance incompetence due to an imbalance of sport-specific and nonsport-specific stressors and recovery with atypical cellular adaptations and responses. Besides performance incompetence, many other clinical problems may arise as a result of overtraining, including sports injuries, infections, or mood disturbances such as fatigue or depression. Imbalance of stress (training-specific, psychological, and non-specific) and recovery determines the outcome of a given training situation. Most clinical problems are observed in training with a high metabolic load of more than 4000 kilocalories per day. Training with lower metabolic demands may also result in performance incompetence and clinical symptoms; however, these problems result mainly from non-metabolic causes rather than sport-specific stressors and incomplete recovery. (pp. 103–104)

Steinacker and Lehmann have provided a definition that describes OT as both a process and an outcome (i.e., OT syndrome), and have included several other possible adverse outcomes associated with the process of OT. Similarly, Hooper and Mackinnon (1995) and O'Toole (1998) outlined many of the possible outcomes of OT, describing OT as a process, or behavior, that leads to a state of nonadaptation associated with negative outcomes, such as prolonged fatigue, depression, illness, injury, and long-term disruption of general physical and psychological well-being (OT syndrome). Hooper and Mackinnon indicated that *overtraining* is a process, whereas *overtraining syndrome* is an outcome, representing the extreme end state of nonadaptation that results from OT behavior.

In contrast to authors who have presented OT in a negative light, Raglin (1993) originally described

OT as an "integral and necessary aspect of endurance training," where it is regarded as a "stimulus consisting of a systematic schedule of progressively intense physical training of a high absolute and relative intensity" (p. 842). Raglin's definition is closely related to the Steinacker and Lehmann (2002) description of normal training (i.e., high-load training, overload training, overreaching, recovery) Hackney, Pearman, and Nowacki (1990) viewed OT as part of the training process, but described it as an abnormal extension that leads to a state of "staleness" or "being overtrained" (p. 22). Kreider, Fry, and O'Toole (1998), however, focused on OT as a state of stress accumulation, without distinguishing OT as a process from OT as an outcome:

> *Overtraining* is an accumulation of training and non-training stress resulting in a long-term decrement in performance capacity with or without related physiological and psychological signs and symptoms of overtraining in which restoration of performance capacity may take from several weeks to months. (p. viii)

If "restoration of performance capacity" takes several months, then one could safely say the athlete is experiencing overtraining syndrome.

Lehmann, Foster, Gastmann, Keizer, and Steinacker (1999) presented OT in both positive and negative terms, distinguishing the definitions by time frame (i.e., short- or long-term):

> *Short-term overtraining* (also called overreaching or supercompensation training) is a common part of athletic training, which leads to a state of overreaching in affected athletes. This state of overreaching is characterized by transient underperformance, which is reversible within a short-term recovery period of one to two weeks and can be rewarded by a state of supercompensation (an increase in performance ability following one to two weeks of regeneration after a short-term phase of overtraining); therefore, short-term overtraining, or overreaching, is a regular part of athletic training. . . . (p. 2)

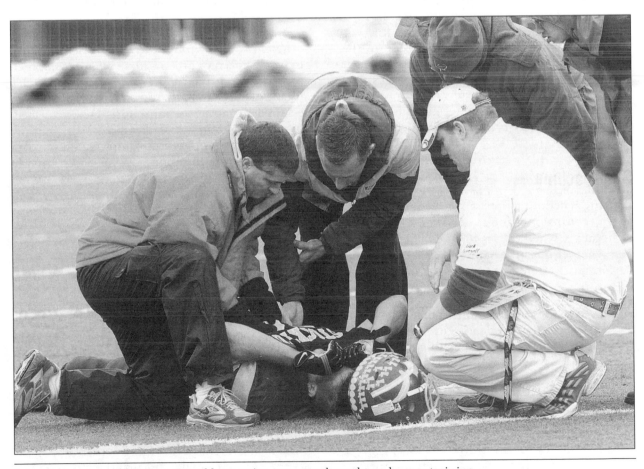

Injury is just one of the many possible negative outcomes brought on by overtraining.

Long-term overtraining occurs when over-reaching is too profound or is extended for too long; this occurs if the necessary regeneration period is inappropriately short and recovery therefore remains incomplete and is additionally associated with too many competitions and non-training stress factors. The athlete clearly runs the risk of a resulting overtraining syndrome. (p. 2)

With these definitions, Lehmann et al. (1999) have still attached positive implications to OT, equating short-term OT with overreaching, and seeing it as a necessary process in achieving optimal performance. Armstrong and VanHeest (2002) noted the ongoing debate about positive and negative usages of the terms *overtraining* and *overreaching*, stating that "some authorities view overreaching as a deliberate attempt to induce optimal performance" and "others view it as an unplanned, undesirable outcome of strenuous training" (p. 187). Armstrong and VanHeest indicated that, although they saw overreaching as positive, defining it in terms of a process that brings about supercompensation, they viewed OT, short- or long-term, as negative and associated with chronic performance decrement. We agree with Armstrong and VanHeest that using a term (i.e., *overtraining*) to stand for both positive and negative aspects of training confuses the issues. Thus in this book we will use the terms *overtraining* and *overreaching* to indicate the negative and positive aspects, respectively, of high-load or high-intensity training.

Overreaching

Looking at definitions of *overreaching,* there is, once again, confusion—about whether it is positive or negative, whether it is to be equated with OT, and whether it is necessary for achieving optimal performance. Here are two definitions from the literature:

Overreaching refers to training that involves a brief period of overload, with inadequate recovery, that exceeds the athlete's adaptive capacity. This process involves a temporary performance decrement lasting from several days to several weeks. (Armstrong & VanHeest, 2002, p. 187)

Overreaching is an accumulation of training and non-training stress resulting in a short-term decrement in performance capacity with or without related physiological and psychological signs and symptoms of overtraining in which restoration of performance capacity may take from several days to several weeks. (Kreider et al., 1998, p. viii)

Although some coaches, researchers, and others might claim that overreaching is necessary during the training process in order to improve performance, the authors of a consensus statement outlined at a Human Performance Summit convened by the U.S. Olympic Committee and the American College of Sports Medicine concluded that overreaching should be avoided because of its unpredictable outcomes (Urhausen & Kindermann, 2002). Such comments questioning the necessity of overreaching may be at odds with many training practices. Following the overload principle (i.e., overreaching) is one of the most effective and accepted ways to improve physical performance. So why did the USOC/ACSM summit caution against overreaching? We do not know, but it may be that the definition they used, like some definitions for overtraining, was multifaceted and inconsistent.

Having been a competitive athlete in several different sports throughout his life, Sean has yet to hear a coach or other athlete refer to a planned process of training as *overreaching* or *overtraining*. Elite athletes train hard, often extremely hard, and get tired during heavy training; fatigue is expected, no matter what it is called. After heavy training, if the training and recovery are well managed, athletes recover and (one hopes) perform well in competition. If the training and recovery are not well handled, then athletes will not recover sufficiently to perform at peak when they need to at competitions.

Athletes and coaches in the real world are applying the overload principle to training in lead-ups to major competitions. We doubt if they are interested in the definitional and semantic arguments of researchers in the field. What they are probably most interested in are the factors that influence whether applying the overload principle results in improved performance, unchanged performance, or deterioration of performance. For coaches and athletes, the labels we give to phenomena or training practices have little relevance to the main object of interest, and that is performance. The question they want answered is: How can we monitor normal and overload training in order to get positive results and avoid damaging the athlete?

Staleness

If there were not already enough confusion about overreaching and OT, researchers often use the term *staleness* to refer to the state of sustained fatigue or underperformance experienced by athletes. Staleness ostensibly involves a less severe stage in the development of full-blown OT syndrome (Silva, 1990). It has been regarded as an undesirable response that is a consequence or product of OT (Raglin, 1993). It has also been equated with OT syndrome (Hooper & Mackinnon, 1995), and Hackney et al. (1990) defined it as

> a state in which the athlete has difficulty maintaining standard training regimens and can no longer achieve previous performance results (i.e., performance decline). The terms "staleness" or "being overtrained" are commonly used interchangeably. This term can be defined as the 'end result.' (p. 22)

Silva (1990) described staleness as "an initial failure of the body's adaptive mechanisms to cope with psychological and physiological stress" (p. 10). Perhaps, like croissants, stale athletes could be described as a little moldy, dry, and crusty around the edges. Although most researchers appear to be in agreement that staleness is an outcome of the OT process, and is pretty much the same thing as OT syndrome, there does not seem to be any good reason to use the term, aside from adding more jargon to an already confusing lexicon in the field. That said, "staleness" might be used to describe a less severe form of OT syndrome (where recovery takes a couple months rather than 6 months or a year), but it seems like splitting hairs, and we will not be using the term in this book.

Burnout

In looking at athletes' changes in performance and struggles with fatigue, researchers have also added the term *burnout* to the list of descriptors associated with OT. After having read much of the literature on burnout in sport, and the original burnout literature in the sphere of human services, we are left with the sense that *burnout* and *overtraining* are terms that can be confused, because they both describe processes surrounding, and responses to, stress overload. Furthermore, *burnout* is a term that we have heard used frequently, among athletes and coaches alike, to describe feelings of being fed up with some, or all, aspects of their sport. Burnout

appears to operate more at the cognitive (McCann, 1995) and emotional level, whereas overtraining has a necessary physiological component (negative effects of hard physical training) that is not usually a part of burnout as originally defined. Looking at the original definitions, however, the most significant distinction seems to be that *burnout* was used to describe stress responses in people working in human services and *overtraining* was used to describe stress responses in athletes training in competitive sports.

Clinical psychologist Herbert Freudenberger (1974) coined the term *burnout* to describe stress responses among staff members of clinical institutions, such as free clinics and halfway houses. Most burnout research has continued to focus on people in human services occupations and has been based on Maslach and Jackson's (1981) definition of *burnout* as a stress reaction syndrome involving three dimensions: emotional exhaustion, depersonalization, and feelings of low personal accomplishment. Maslach (1982) suggested that burnout

> is a response to the chronic emotional strain of dealing extensively with other human beings, particularly when they are troubled or having problems. Thus, it can be considered one type of job stress. Although it has some of the same deleterious effects as other stress responses, what is unique about burnout is that the stress arises from the *social* interaction between helper and recipient. (p. 3)

According to Maslach's description, it seems that, as with overtraining, burnout is about experiencing an overload and reacting negatively to it. Nonetheless, Maslach makes clear that burnout is unique and distinguishable from other experiences of stress, in that it is about *job stress*, and, in particular, stress arising from human services employment. This issue regarding psychosocial job stress seems to be the clearest point of departure from similarities to OT in athletes, which would not be described as a stress response arising from a helper–recipient relationship.

Athletes may overtrain physically while still being psychologically motivated, but they cannot overtrain without the presence of the physical stressor. In contrast, burnout seems to involve physical and psychological stressors, together or independent of one another, but it always includes the cognitive interpretation of the stressors. Athletes may become burned out from psychological stressors alone or from a combination of psychological and physical stressors.

It appears that *burnout* shares similarities in signs and symptoms with OT. Burnout sounds like an appealing term to apply to athletes because it describes how people get exhausted in response to chronic stress. Similarly, overtrained athletes can be characterized as being exhausted in response to chronic stressors. The loss of motivation inherent in burnout, however, seems to be a key factor in distinguishing it from OT; athletes who are overtraining often maintain very high levels of motivation. Burnout may be considered one outcome of the OT process. Overtrained athletes may get fed up with stress overloads and feel like they don't want to do their sports anymore, but burnout can exist independently of OT as well. Athletes may get tired of training, lose motivation, and feel emotionally exhausted but not suffer the physical breakdowns associated with OT. Given the original definitions of *burnout*, the applications of the term in sport, and previous discussions of OT terminology, there does not seem to be much benefit to using the term in association with OT. Nevertheless, studies about athletes becoming unmotivated, sick of their training, and apathetic (i.e., burned out), even though not physically overtrained, would be beneficial to coaches and athletes alike.

Underrecovery

Finally, in an attempt to shift attention away from training, OT, and the confusion surrounding these terms, Kellmann (2002) has moved to a focus on the recovery aspect of athletic experience. According to Kellmann's view, instead of OT, athletes can be described as experiencing "underrecovery":

> *Underrecovery* is the failure to fulfill current recovery demands. Underrecovery can be the result of excessively prolonged and/or intense exercise, stressful competition, or other stressors. Underrecovery can result from training mistakes, such as monotonous training programs, more than three hours of training per day, more than a 30 percent increase in training load each week, ignoring the training principle of alternating hard and easy training days or by following two hard days with an easy day, no training periodization and respective regeneration micro-cycles after two or three weeks of training, or no rest days. (p. 3)

The introduction of the term *underrecovery* might draw attention away from the "over" terminology and take the heat off of some coaches for training athletes too intensively. Kellmann even suggested that "insufficient and/or lack of recovery time between practice sessions is the main cause of the overtraining syndrome" (p. 12). Budgett (1998) supported this statement, proposing that underrecovery, not necessarily an excess of training, leads to the overtraining syndrome.

The shift in focus from training to recovery might also be useful in emphasizing neglected areas of athletes' lives, as well as for highlighting individual differences in recovery needs. Kellmann (2002) defined *recovery* as an "inter- and intra-individual multilevel (e.g., psychological, physiological, social) process . . . for the reestablishment of performance abilities" (p. 10). It involves physiological factors (e.g., restoring nutritional resources, getting sufficient sleep), psychological components (e.g., feelings of relaxation, sense of well-being, positive moods), and social activities (e.g., gatherings with friends, healthy relationships with others). Looking at these aspects of recovery broadens the focus on what might be leading athletes to feel fatigued, beyond just the volume and intensity of the training program. Furthermore, as Kellmann noted, "recovery is specific to the individual and depends on individual appraisals" (p. 7). Each athlete has recovery strategies, and what works for one might not work for another. Kellmann referred to an individual athlete's response to training and non-training stressors as the *recovery-stress state*, a term that encompasses Lehman et al.'s (1993) earlier description of factors that threaten to imbalance athletes' lives. Kellmann stated:

> Inter-individual differences in recovery potential, exercise capacity, non-training stressors, and stress tolerance may explain the different degrees of vulnerability experienced by athletes under identical training conditions. (p. 16)

Kellmann concluded his discussion of underrecovery by suggesting that it has the same impact as OT, in that performance declines, but he noted that underrecovery is the precursor or cause of the OT syndrome. Using the term *underrecovery* seems to have highlighted some important factors on which coaches, athletes, researchers, and others might focus to ensure a holistic and healthy approach to athletic training. Nonetheless, the concept of balancing stress and recovery behaviors may still elude encapsulation by a single term. Underrecovery phenomena are highly variable. Each athlete responds differently to physical training and psychosocial stressors. Even for a single athlete, the same physical training may be too much or just right depending on the other stressors operating at that time. There is no formula for predicting under-recovery, and only careful and constant monitoring of athletes' training and psychosocial demands and stressors will help coaches detect when athletes are exceeding their limits (see chapter 11 for more suggestions).

The Latest Word(s)

Most recently, several new combinations of terms have been used in the overtraining literature (e.g., Meeusen, 2006; Nederhof, Lemmink, Visscher, Meeusen, & Mulder, 2006; Uusitalo, 2006). Applying the overload principle successfully (high-load or high-intensity training leading to performance decrements, but with a positive response to taper) is called *functional overreaching*. Unsuccessful overloading (absence of performance rebound after taper) is called *nonfunctional overreaching*, and severe cases where performance does not improve for a long period of time are labeled with the tried-and-true term *overtraining syndrome*. Whether these terms will take hold with overtraining researchers, only time will tell.

WHY LANGUAGE IS IMPORTANT

Labels and definitions may be helpful in keeping an area of research focused. It seems, however, that the various definitions of OT processes and outcomes have created some confusion, or at least disagreement, among researchers and others. Perhaps the confusion and disagreement have arisen because the process of balancing stressors and different forms of recovery, life issues, and intra- and interpsychic challenges is not easy to describe with a few concise terms. Generally, in OT research, there has been a focus on one outcome: OT syndrome. Many of the definitions presented, whether describing OT processes or outcomes, exclude reference to other significant outcomes, such as injury and illness, which some researchers have related to the same underlying behaviors.

Furthermore, based on our collective experiences as competitive athletes and practitioners, it seems that many coaches and athletes in North America, the United Kingdom, and Australia interpret any usage of terms such as *overtraining*,

overreaching, or *over* anything, as negative. Coaches and athletes may not be happy to say that they are engaging athletes in *short-term overtraining* or *overreaching*, even if researchers are using these terms to describe effective forms of training. Coaches seem particularly defensive about using any such terms, because the implication for them is that there is something wrong with their training programs. For coaches and athletes, there is just "training," and either athletes are training well or they are not.

It seems pragmatic, then, to look at how competitive athletes and coaches work with these terms. As mentioned, in sport, there is training, which is either effective or ineffective in bringing about improved performance. And there is an assumption that training has to be pushed, or increased, as fitness and performance improve, in order to stimulate further gains. Armstrong and VanHeest (2002) referred to this approach as *overload training*, "a planned, systematic, progressive increase in training stimuli that is required for improvements in strength, power and endurance" (p. 187). Training programs are already based on principles that demand increases in training that challenge the body. To use the terms *overtraining* or *overreaching* to describe this process of challenging the body with increased loads seems superfluous, and can be confusing to many, given that the use of "over" as a prefix implies that one has done something excessively. To talk about positive aspects of OT, therefore, seems contradictory. On the one hand, there is effective training, and, on the other hand, if OT is involved, whether short- or long-term, there is ineffective training. The sticking point seems to come from the original usage of the term to describe the process of overload training that is required for optimal performance. This usage may have reflected a shift in the way athletic training was approached when overloading (as a principle) was first taken on board by athletes, coaches, and sport scientists. It seemed that there was a need to differentiate regular training, which now might be seen as maintenance training or undertraining, from the more effective form of overload training.

STRESS AND RECOVERY: TAKING A BROADER VIEWPOINT

Perhaps a general and useful approach to understanding athletes' (and coaches') experiences with OT processes and outcomes is to stay focused on

HOW BIG IS IT?

The Prevalence and Manifestation of Overtraining

In chapter 1, we described the broad and narrow definitions of overtraining (OT) and some of the deleterious effects that can occur, but just how big is the problem? We appreciate how devastating it must be for individual athletes to overdo it in the build-up to the most important competitions of their lives. Loss of a key player due to overtraining at a crucial stage in a competition can be equally dispiriting for a whole team. For most of us, the main sign that an athlete has overtrained is sustained poor performance in competition. But poor performance can also be accounted for by many other factors, such as inadequate preparation, loss of confidence, or pressures outside of sport. So, how do we know when athletes have overtrained? What are the characteristics of overtraining? And how often does it happen?

In this chapter, we move from the history and language of overtraining to some of the ways in which researchers have tried to understand and measure it in order to determine whether it has occurred or, preferably, if it is happening, so that interventions can be initiated. We have proceeded thus far on the assumption that overtraining is an important phenomenon to study because our intuition and experiences tell us that it happens often, and at a level that has powerful personal and team effects, but its true prevalence has been an important question for researchers. Of course, this issue is closely linked to our conceptualization and measurement of the phenomenon. To determine how often overtraining occurs, researchers must decide what they mean by it. Then they have to find ways to operationalize the concept; that is,

they have to measure it. Whereas some researchers have employed the kind of qualitative methods of inquiry that we advocate in this book, much research on the prevalence of overtraining is based on physiological, psychometric, and behavioral measurement techniques.

The waters seem to have been muddied by an issue related to conceptualization and measurement: the distinction between processes (e.g., what goes on to produce overtraining) and outcomes (e.g., injury, illness, overtraining syndrome). As we have written in chapter 1, this muddiness is reflected in the language of overtraining, where the term *overtraining* is usually employed to refer to the process; an athlete *is* overtraining. When we talk about the outcomes, we commonly refer to "overtraining syndrome"; a performer *has* overtraining syndrome.

Another important conceptual issue is the relationship between overtraining, injury, and illness. Historically, researchers who have examined overtraining generally have not fully integrated injury or illness as outcomes (see all entries in the reference section for this chapter with Morgan as the first author for examples of studies that did not integrate injury and illness as outcomes). There are exceptions. For example, Fry, Morton, and Keast (1991) and Kuipers (1996) have discussed mechanical overtraining and overuse injuries along with metabolic overtraining (or overreaching) and immune system problems.

Recently, some researchers have started to think about ways in which the physical and psychological stressors that lead to overtraining are difficult to distinguish from those that result in many injuries and a range of illnesses (e.g., Steinacker & Lehmann, 2002). We also suggest that overtraining should be considered as part of a broad process through which a range of physical, psychological, and social stressors—with which athletes live every day—can contribute to injury, illness, and overtraining.

To provide a foundation for exploring the overtraining phenomenon, we discuss in this chapter the prevalence of overtraining, the measures or markers of overtraining, and the outcomes of overtraining (i.e., the links between overtraining and injury and illness).

PREVALENCE RESEARCH AND DATA FOR OVERTRAINING

The prevalence of OT syndrome in specific sports has not yet been clearly established. The literature does include regular citations of a number of studies with some prevalence data for OT (Hooper, Mackinnon, Howard, Gordon, & Bachmann 1995; Morgan, Brown, Raglin, O'Connor, & Ellickson, 1987; Morgan, O'Connor, Ellickson, & Bradley, 1988; Morgan, O'Connor, Sparling, & Pate, 1987). Many of these studies, however, have weaknesses in that the statistics were often based on unclear classifications of OT and OT syndrome, or on a very small number of athletes. In two studies examining characteristics of elite distance runners, Morgan, O'Connor, et al. (1988) and Morgan, O'Connor, et al. (1987) reported that 64% of males and 60% of females, respectively, indicated they had experienced *staleness* (Morgan, O'Connor, et al.'s (1988) synonym for *OT syndrome*) at some point during their careers. In both studies, researchers used the same definitions of OT syndrome to gather frequency data. Morgan, O'Connor, et al. (1988) suggested that staleness is "usually characterized by a variety of behavioral, psychometric, and physiologic symptoms with perhaps the most salient features being:

> performance decrements or the inability to train at customary levels,
>
> chronic fatigue, and
>
> depression of clinical significance." (p. 251)

Looking at this description of staleness, or OT syndrome, we are not sure if a symptom of OT syndrome should be described as "psychometric." A symptom is not a metric. Furthermore, from what

was reported in the studies, there did not seem to be any clinical assessment of depression in the athlete participants, although this psychological disorder was deemed to be an important component of the definition of staleness or OT syndrome. Depression is a clinical syndrome itself, with specific criteria. The researchers also did not state whether the concept of staleness was explained or described to athletes before they were asked whether they had experienced staleness during their careers. And given the unclear description of staleness in the publications, it seems that the athletes might not have been in a position to make reliable statements about their OT experiences even if the researchers did explain the concept to them. In addition, memories are fallible, and, in reflecting on one's entire career, there have to be periods of performance decrements (no one gets better all the time), periods of being tired for weeks, and times when one has been sad. It seems doubtful that 60% of the athletes had clinical, or even subclinical, depression. These problems leave the reported frequencies suspect.

In a series of studies spanning 10 years, Morgan, O'Connor, et al. (1988) monitored mood states, using the Profile of Mood States (POMS; McNair, Lorr, & Droppleman, 1992), in college swimmers, and suggested that during peak, twice-daily training, "it is not uncommon for 5–10% of the swimmers to experience what we regard as staleness" (p. 108). Although Morgan, O'Connor, Ellickson, et al. may have developed sensitivities for which swimmers were overtraining during their 10 years of observing such athletes, the prevalence statistics presented were only from an estimate made in passing, not from a frequency value derived from systematic inquiry.

In contrast to these studies where athletes were classified as stale or overtrained based on general and sometimes vague definitions of OT syndrome, Hooper et al. (1995) reported an OT syndrome frequency statistic in a small group of Australian swimmers using a specific classification of OT syndrome, which they also referred to as *staleness*. Hooper et al. stated that swimmers were classified as stale if all of the following occurred:

> (a) failure of performance in the maximal effort swim to improve from early- to late-season; (b) failure of performance in the trials to improve from previous best times; (c) fatigue ratings in the [training] logs > 5 (scale 1–7) for more than 7 days consecutively; (d) comments in the page provided

in each log that the athlete was feeling as though he or she was responding poorly to training; (e) a negative response to a question regarding presence of illness in the swimmer's log, together with normal leukocyte count and ESR [erythrocyte sedimentation rate] at testing time. (p. 108)

Based on these specific classification criteria for staleness or OT syndrome, Hooper et al. found 3 out of the 14 swimmers (approximately 21%) could be identified as stale by the end of the season, that is, following the Australian national swim titles. Here, Hooper et al. provided some idea of prevalence of OT syndrome, based on clear, specific criteria, among a group of elite swimmers. Unfortunately, the small sample size makes it difficult to generalize the finding to athletes in other sports.

In examining much larger samples of athletes, Gould, Greenleaf, Chung, and Guinan (2002) noted that 28% of 296 U.S. Olympians at the Atlanta games (1996), and 10 percent of 83 U.S. Olympians at Nagano (1998), reported that they were overtrained in the 90 days prior to the Games, and that the overtraining had a negative effect on performance. In all cases, Olympians in the

Gould et al. studies were identified as overtrained if they answered *yes* to the statement "I overtrained in preparation for the Olympics" (p. 181). The researchers, however, appear not to have provided either the Atlanta group or the Nagano group with definitions or explanations of OT. Given the concerns already discussed about clearly defining OT, it would seem that these prevalence statistics are not reliable; self-reports of OT, without any reference to OT criteria or definitions, do not seem to provide much useful data. Also, one might consider that athletes who were not as successful at the Games as they hoped to be might respond to the question in a self-protective manner that puts blame on their training rather than on something more personal.

In another investigation, looking at the problem of overtraining in adolescent athletes, Raglin, Sawamura, Alexiou, Hassmén, and Kenttä (2000) asked 231 young swimmers from Greece, Japan, Sweden, and the United States if they had ever experienced a loss of performance, for at least 2 weeks, that resulted not from injury or illness but from training. Raglin et al. used this definition (i.e., perceived loss of performance) as the criterion for classifying athletes as having experienced OT, which the researchers referred to as *staleness*.

In a series of studies spanning 10 years, researchers found that during periods of high training, up to 10% of college swimmers may experience states of sustained fatigue or underperformance.

Across the four countries, an average of 35% of the young swimmers reported experiencing the perceived performance losses, 20% from Sweden, 24% from the United States, 34% from Japan, and 45% from Greece. Subsequently, the researchers classified these 35% as having experienced staleness at some point during their swimming careers. Although Raglin et al. employed a more specific question than simply asking whether athletes had overtrained, it is not clear how reliable athletes are at retrospectively assessing episodes of performance loss. How did these athletes decide what constituted a loss of performance? Any number of variables could affect their perceptions. For example, perhaps some did not do as well as they hoped at certain swim meets but were physically healthy, or perhaps comparing themselves with others gave them the impression that they were not performing well for a period of time. Maybe other young swimmers felt pressure from parents that made them feel they were not doing well, even though, objectively, they might have been performing adequately. In these examples, one would probably not want to attribute any perceived performance decrement to OT. Retrospective recall, especially spanning a whole sport career, is a dubious way to gather performance data. Asking athletes to refer to objective markers of performance in answering questions about performance decrements might help to produce more valid data, especially if augmented with corroborating evidence from coaches and others, such as competition results, training results, or objectively identified episodes of performance decrement. Also, as established in the previous chapter, overload training with proper recovery can increase performance at a given competition, if recovery takes place at the correct time. These feelings of performance loss could have occurred directly before recovery and, therefore, been viewed as a decrease in performance even as the athlete was headed toward an increase.

Reflecting on the prevalence research, it is clearly difficult to define and distinguish between overtraining and overtraining syndrome in the first place, and it follows that reliable prevalence data for these phenomena may be hard to obtain. It might be acceptable to speculate that perceived performance decrements stand as preliminary evidence of OT syndrome prevalence, especially when athlete reports of performance losses are augmented with objective measures. Nonetheless, it might still be too big a leap to label such data on performance decrement as clear evidence of OT syndrome frequency in athlete populations.

Summarizing the prevalence research reported here, given very specific criteria for judging OT syndrome, or what many researchers have called *staleness*, it seems that a small but significant percentage of athletes may be classified as having experienced the syndrome. Overtraining not to the point of full-blown overtraining syndrome but to the extent of damaging tissues (e.g., tendons, joints, muscles) or systems (e.g., immune function) appears to happen relatively often and at inopportune times (e.g., the Olympics). The epidemiological research base for this sub-OT syndrome does not exist, and this common tissue- and system-damaging overtraining is poorly understood. We have, however, in our personal sporting careers and professional lives, seen and heard about innumerable cases of athletes training to the point of serious damage. Although research is inconclusive due to many limitations, we can conclude that there are probably quite a few athletes who are experiencing, or have experienced, various levels of overtraining at some point in their careers.

We felt that we needed to understand both the limited phenomenon of OT syndrome and the more frequent occurrence of system- and tissue-damaging training. Given the state of the research and current data available, in-depth qualitative research on athletes', coaches', and exercise scientists' experiences with these two types of processes and outcomes of training seemed to be the best way to develop a comprehensive picture of these phenomena. That approach forms the core of this book.

MARKERS OF OVERTRAINING

Although underperformance is regarded as the hallmark of OT syndrome, it is not clear how much performance has to drop to indicate a state of OT. Performance decrements may be the result of OT or of other precipitating factors (e.g., family problems)(Hooper & Mackinnon, 1995; O'Connor, 1998; Raglin, 1993). Athletes, coaches, and sport scientists have been interested in finding valid early warning signals upon which they can act to prevent undesired underperformance. In this section, we briefly list the research on physiological markers of overtraining, then examine psychological markers. We must acknowledge that our expertise is in psychology, and we will devote the largest part of this section to psychological markers. Readers interested in physiological markers might also refer to the reference section for this chapter and the list of suggested readings at the end of this book.

Physiological Markers of Overtraining

Much of the research has focused on assessing the onset of OT syndrome, and, although no single marker or group of markers has been identified, the following physiological factors have been demonstrated to have significant associations with OT, where OT was determined by performance decrement:

- hormonal responses to exercise load (plasma adrenaline, serum cortisol; Uusitalo, Huttunen, Hanin, Uusitalo, & Rusko, 1998),
- changes in ratio of free testosterone to cortisol (FTCR; Chicharro et al., 1998),
- maximal lactate concentrations during incremental graded exercise (Jeukendrup & Hesselink, 1994; Urhausen, Gabriel, Weiler, & Kindermann, 1998),
- hypothalamo-pituitary dysregulation (Urhausen, Gabriel, & Kindermann, 1998),
- lowered urinary norepinephrine (Mackinnon, Hooper, Jones, Gordon, & Bachmann, 1997),
- changes in plasma glutamine (Rowbottom, Keast, Garciawebb, & Morton, 1997),
- deterioration in neuromuscular excitability (Lehmann, Baur, Netzer, & Gastmann, 1997),
- decreases in secretory immunoglobulin (IgA; Mackinnon & Hooper, 1994),
- decreased heart rate variability during orthostatic challenge (Uusitalo, 2001),
- decreased heart rate during maximal exercise (Hedelin, Kenttä, Wiklund, Bjerle, & Henriksson-Larsén, 2000),
- decreased muscle glycogen levels (Snyder, 1998), and
- reduced sleep efficiency as measured by wrist actigraph (Wall, Mattacola, & Levenstein, 2003).

Aerobic and Anaerobic OT

There have been some suggestions in the literature that OT can be broken into sympathetic and parasympathetic classifications (Kellmann, 2002; Lehmann et al., 1998; Lehmann, Foster, & Keul, 1993). Sympathetic OT has been associated with such characteristics as increased resting heart rate and blood pressure, decreased appetite, loss of body mass, disturbed sleep, and irritability. Parasympathetic OT has been associated with such patterns

Even though sympathetic overtraining has been linked to power and speed sports, like sprinting, and parasympathetic OT has been associated with endurance sports, like rowing and cross country skiing, the aerobic and anaerobic distinction may not hold up well.

as low resting heart rate and blood pressure, long periods of sleep, and depression (Kellmann, 2002; Lehmann et al., 1993; Mackinnon & Hooper, 1994). Sympathetic OT has been linked to power and speed sports, and parasympathetic OT has been associated with endurance sports (Kellmann, 2002; Lehmann, Dickhuth, & Gendrisch, 1991; van Borselen, Vos, & Fry, 1992). The distinctions between sympathetic and parasympathetic classifications appear clean and make intuitive sense: If an athlete overloads the body in distinctly different ways (aerobic versus anaerobic), but if no physiological marker or group of markers has been unequivocally identified for the OT syndrome, and if none of the changes in physiological markers can differentiate intensive but effective training from OT, then how can such markers be used to classify different types of OT? Based on the research findings, it is not clear whether such distinctions add clarity or increase confusion in trying to understand OT processes and outcomes. Fry (1998) argued however, that aerobic and anaerobic exercises should be considered differently in the context of overtraining:

> It is quite evident that the adaptations to aerobic, or endurance, types of exercise are quite different from the adaptations to anaerobic exercise . . . [such as] resistance exercise. . . . What has not been as apparent in the overtraining literature is that overtraining with endurance exercise is also quite different from overtraining with resistance exercise. . . . As a result, one must be wary of using the endurance-overtraining literature to infer what happens during overtraining with resistance exercise. (p. 110)

We are not sure that the aerobic versus anaerobic distinction holds up well. Endurance athletes (e.g., rowers) engage in significant resistance training, and anaerobic sprint athletes (e.g., 50 and 100 meter swimmers) engage in substantial endurance training. Trying to determine which kinds of training may result in which kinds of OT outcomes may be conceptually appealing, but not practically useful given the different constellations of aerobic and anaerobic training in most sports.

In a study looking at increased volume training with U.S. national-level judo athletes over 6 weeks, Callister, Callister, Fleck, and Dudley (1990) suggested that intensive anaerobic training may present a different set of symptoms from those displayed in endurance overtraining. Nonetheless, with increased volumes of anaerobic training and concomitant decreases in performance, expected changes in sympathetic nervous system activity did not occur; and there were no significant increases in resting heart rate and blood pressure.

In a pilot study on intensive resistance (anaerobic) training, Fry, Kraemer, Lynch, Triplett, and Koziris (1994) found no decrease in performance for maximal strength following an overload protocol, but did observe decreases on other physical tasks, such as speed-controlled strength and sprint speed. The investigators concluded that an overload stimulus in resistance training might create an OT response in non-training-specific musculature, which is important for athletes and coaches to consider when designing training programs. To gain greater insight into the types of overtraining associated with anaerobic overloads, Fry, Kraemer,

van Borselen, et al. (1994) conducted a follow-up study, during which they tested participants' responses to an even more intensive resistance training protocol than the one from the pilot study. In the follow-up study, the researchers also looked at endocrine adaptations to the training. With higher intensities of resistance training, all participants experienced both decreased maximal strength on the training-specific task and decreased performance on non-training-specific tasks. To differentiate this type of overtraining from endurance types of overtraining, the investigators noted that participants did not display symptoms normally associated with aerobic overtraining (i.e., changes in sleep patterns, resting heart rate, or body composition). Furthermore, Fry (1998) later noted that endocrine profiles of these athletes were quite different from those associated with athletes training under endurance overloads:

> With high relative intensity resistance exercise overtraining, resting concentrations of both epinephrine and norepinephrine were unaffected, but acute concentrations exhibited considerable increases. . . . In general, the elevated catecholamine response to the resistance exercise stimulus during high relative intensity resistance exercise overtraining is evidence of the sympathetic overtraining syndrome. . . . It is also quite different from the attenuated catecholamine levels (i.e., parasympathetic overtraining) reported for overtrained endurance athletes, again indicating the unique differences of training and overtraining with different modalities and protocols. (p. 118)

Fry (1998) also suggested that elevated catecholamine levels, together with attenuated muscular performance, may be indicative of the onset of a sympathetic OT syndrome. Yet he also pointed out that resistance-trained participants who did not overtrain exhibited significant positive relationships between immediate postexercise circulating concentrations of catecholamines and muscular strength performance. Therefore, in the cases of both OT and non-OT outcomes, there may be elevations in catecholamines, but in OT there is a decrease in performance with no significant associations to levels of circulating catecholamines. It sounds like catecholamine levels are not predictive of performance decrements. Rather, such levels may be associated with high training loads. As with many endurance-training athletes, it seems that when resistance-training athletes increase intensities, they experience physiological changes, which seem to be specific to the type of activities in which they are involved. Such physiological changes occur whether one is over-training or not, and there is no clear boundary that may be indicated by the changes. Thus the sympathetic and parasympathetic distinctions seem to be more about describing differences in response to different types of heavy training that could lead to OT than about classifying or predicting different types of OT syndrome. As with aerobically trained athletes, anaerobically trained athletes might display the same array of physiological signs and symptoms while engaging in effective training that they would while engaging in OT. So it may be best not to worry about distinguishing OT in terms of different outcomes for aerobic or anaerobic activities; rather, it is significant simply to acknowledge that there *are* outcomes associated with excessive overload that affect performance across these different types of sports. Predicting the outcome for a specific athlete, therefore, seems unlikely, no matter what.

The key element for any type of OT syndrome seems to be performance decrement. Although OT athletes might display a number of physiological signs and symptoms, there are no clear physiological markers of OT syndrome, whether athletes are training aerobically or anaerobically. Nonetheless, the research looking at different types of training in the OT context appears to have been important in illustrating that it is possible to overtrain (as defined by sustained performance decrement) in different ways and with different outcomes. Athletes may feel heavy or fatigued and want to sleep a lot, or they may seem jittery and agitated, but in all cases of overtraining there will be clear indicators of sustained performance decrement or stagnation, which do not improve even after a substantial period of recovery.

Despite a bewildering list of potential variables and a substantial body of research on physiological

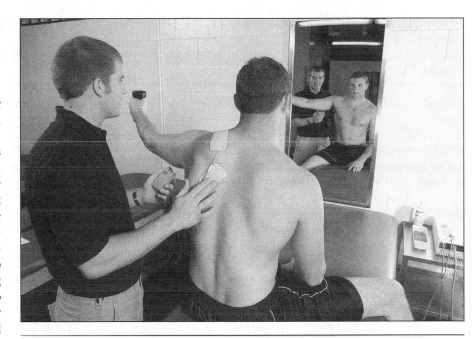

Despite the large body of research on physiological markers of overtraining, reliable early warning signs have yet to be conclusively identified.

markers, reliable early warning signals that athletes are starting to overtrain have yet to be conclusively identified. The literature also includes attempts to link physiological markers with psychological markers of OT, and we discuss these efforts in the next section.

Psychological Markers of Overtraining

The psychological research on OT has added to the physiological perspective, focusing primarily on the relationships between mood (as measured by the Profile of Mood States, or POMS), training load, and subjective ratings of well-being. Researchers have suggested that increased mood disturbances and self-reports of decreased well-being may be valuable indicators of impending OT syndrome (Berglund & Säfström, 1994; Fry, Kraemer et al., 1994; Hooper et al., 1995; Hooper, Mackinnon, & Hanrahan, 1997; Morgan, Costill, Flynn, Raglin, & O'Connor, 1988). Given that the research on psychological variables in overtraining has focused almost exclusively on the POMS, we will now review some of the more salient research in this vein.

Much of the POMS research has evolved from a series of studies by Morgan, Brown, et al. (1987) that consider mood states in collegiate swimmers across a 10-year period. The researchers monitored mood states using the POMS across the competitive seasons for groups of male and female swimmers and reported that mood state disturbances increased in a dose–response manner in accordance with training loads. With increased training loads, POMS global mood disturbance scores increased, as measured by an aggregate of all the subscales (with the *vigor* subscale negatively weighted). Morgan, Brown, et al. suggested that "monitoring of mood states during a given macro-cycle offers a potential method of quantifying distress and titrating training loads on an individual basis" (p. 113). It appears that Morgan, Brown, et al. took significant steps to initiate research on the relationships between mood states and training loads. Some issues, however, stand out from this research and have been carried through much of the subsequent investigations of the POMS in sport performance and overtraining. From the series of studies that Morgan, Brown, et al. conducted, it appears that something is happening with POMS scores in relation to training load. Generally, as training loads increase, scores on the negative POMS subscales of *tension-anxiety*, *depression-dejection*, *anger-hostility*, *fatigue*, and *confusion-bewilderment* tend to increase, whereas scores on *vigor* tend to decrease. Nonetheless, it is not clear what POMS scores reveal about individual athletes or how such scores might be most useful to coaches and athletes. Morgan, Brown, et al. provided individual profiles for only 2 of approximately 400 swimmers. The researchers did not provide performance data or demonstrate any links between POMS profiles and individual performance. Using group norms to suggest what is going on with the POMS and athletic training obscures what is happening at the individual level. Reading this research, we do not have any idea what an individual POMS profile might tell us about any particular athlete. The authors emphasized that athletes who became stale displayed high levels of mood disturbance, but they did not indicate whether there were athletes who became stale without the corresponding mood disturbances, or if there were athletes with high mood disturbances who did not become stale. Therefore, with no links to performance, and with observations reported only on a group basis, coaches looking at individual POMS profiles would not be certain about whether they were observing athletes who were coping well with training or who were not.

Morgan, Costill, et al. (1988) conducted a study with swimmers that looked at effects of short-term (12-day), high-intensity training on mood states. The researchers found that for the group as a whole mood disturbance increased from day 1 to day 5, then remained elevated for the following 7

days. The increases, however, seemed to be mostly attributable to three out of the nine swimmers who displayed markedly higher profiles on average than the rest of the group (i.e., these three swimmers elevated the group mean for mood disturbance; the other six had only mild disturbances). These three swimmers were also found to have significantly lower muscle glycogen levels and were not able to tolerate the prescribed training load. The authors reported that these three swimmers were not ingesting enough carbohydrate and were training at a caloric deficit of up to 1,000 calories less than the other swimmers. In this study there were no correlations reported between POMS scores and performance parameters. Thus, from what we can gather from the results, the simplest explanation for the increases in POMS scores and the inability to tolerate training loads seems to be that these three swimmers were not eating enough.

Murphy, Fleck, Dudley, and Callister (1990) followed 15 judo athletes during a 10-week training cycle at the U.S. Olympic Training Center, monitoring their mood states for different phases of high-volume and high-intensity training. The researchers found no significant changes in total mood disturbance scores. They did report significant increases on the fatigue subscale scores after week 8, compared with baseline measures. They also reported a significant increase on the anger subscale between baseline measures and week 10 after intense training. Performance on strength and anaerobic endurance tasks declined during the study, but the authors were not able to measure judo performance per se. No performance parameters were linked to POMS scores. The authors commented that for these athletes "the typical signs of overtraining reported in the literature were not observed in response to a 50% increase in conditioning training volume" (p. 48). The authors also stated, however, that "the increased sport specific training had no demonstrable beneficial effects on psychological or performance measures," other than athletes reporting they felt "close to their peaks" (p. 48).

Verde, Thomas, and Shephard (1992) administered the POMS to 10 highly trained male distance runners on three occasions over 9 weeks of training, involving a period of deliberately increased training. The researchers reported significant increases in total mood disturbance during the increased training phase, with subsequent reductions in mood disturbance during the recovery phase. The researchers stated that none of the runners could be classified as overtrained during the study. Nonetheless, they noted 6 of the 10 reported symptoms indicating they were close to a threshold of excessive training, "which most coaches would like to detect" (p. 173). As with other studies, no links were tested or illustrated between POMS scores and performance parameters; nor were individual POMS score patterns discussed in relation to performance. Once again, then, one is left with a vague picture of what POMS scores might reveal in the context of overtraining. Our central criticism of many of these studies is that if POMS scores are not related to the real outcomes of interest (e.g., performance, illness), then any changes in POMS during training periods are relatively meaningless.

Berglund and Säfström (1994) were among the first researchers to use individual POMS scores in a meaningful way to modulate training. With a group of canoeists training for the Olympics, the researchers used POMS scores to titrate training loads. If athletes scored more than 50% higher than their own baseline off-season global mood disturbance, training was reduced. For athletes who scored lower than their own off-season global POMS plus 10%, training was increased. None of the athletes developed signs of OT in connection with their Olympic training, and the researchers concluded that "the monitoring of psychological mood disturbances is useful in reducing the risk of staleness in canoeists undergoing hard training" (p. 1036).

From an athlete's perspective, we would say that any attempt to give some control to the athletes in terms of making decisions about changes in training volume according to how they feel could be useful. From our experience, it seems that many athletes will not speak up about how they are tolerating a given training program for fear of how coaches and other athletes might react. It appears unique to set up a model where training is reduced or increased according to how athletes report feeling. Such an approach changes the whole dynamic of the training environment, which all too often leaves athletes feeling pressured into doing more than they can handle at certain times, and perhaps responding by doing less than what is optimal at other times. From an applied perspective, this use of the POMS to modulate training appeared to be beneficial to the training outcomes of the athletes.

Nonetheless, from a research perspective, the lack of a control group, and the lack of objective links made between POMS measures and performance parameters, leaves one to wonder if the POMS was useful in its own right or as an indicator of overtraining. It is unclear whether these athletes might have had the same results, or better ones, even without the training changes made in accordance with POMS scores. We know a number of coaches who might read the results of this study in a very different light, saying that the athletes probably would have done even better if their training had not been reduced at any time (although such coaches would probably have agreed with all training increases).

In one of the first studies to examine statistically the link between POMS scores and performance after intense training, Hooper, Mackinnon, and Hanrahan (1997) found no significant correlations between POMS scores and maximal effort swimming performance for a group of nationally ranked swimmers (8 males and 11 females). Looking at the pattern of POMS scores across a 6-month season, the researchers also found no significant changes in any of the POMS measures with tapering. They did classify 3 swimmers in their study as stale, according to strict criteria:

> failure of performance in the maximal effort swim to improve from early- to late-season testing; failure of competitive performance in the Winter Nationals to improve from performances in the previous Summer Nationals; log book fatigue ratings above 5 (scale = 1 to 7) for a minimum of 7 consecutive days during the month prior to late-season testing; comments [in the log book] . . . that the athlete was feeling as though he/she was responding poorly to training; and a lack of illness during testing and competition. (p. 4)

Only two of these three stale swimmers displayed elevated mood disturbance profiles. Furthermore, those two reported such disturbances during the early season, tapering, and postcompetition, but not during the high-intensity, high-volume cycles of training, suggesting that their mood disturbance was probably not linked to excessive training loads. Hooper et al. stated that "the data suggest the POMS may not be a sensitive indicator of staleness under all circumstances and may not necessarily differentiate between [on one hand] stale and [on the other hand] intensely trained, but not stale, athletes" (p. 9). They also noted that "the decrease in physiological stress of training with tapering may have coincided with an increase in psychological stress with impending competition, making changes in mood states unlikely until after competition ceased" (p. 10).

Their observations suggest that POMS scores may be affected by many kinds of variables. Some athletes show elevated scores with intense training that is effective, and some show elevated scores with intense training that is excessive and might lead to overtraining syndrome. Some show elevated scores because of the stress of upcoming competition, and some show elevated scores for reasons completely unrelated to training or competition, whereas others do not show any corresponding elevations of POMS scores in any of those situations. It seems, therefore, that POMS scores are not reliable markers of OT. Hooper et al. commented that "while it appears that the POMS may be useful for monitoring those athletes predisposed to staleness, it may not reliably differentiate between stale and non-stale athletes under all circumstances" (p. 11). There also remains the question of how to classify an athlete as predisposed to overtraining in the first place.

In another study testing the relationship between POMS scores and performance after intense training, Hooper, Mackinnon, and Howard (1999) reported significant correlations between POMS confusion subscale scores and changes in maximal swim effort after a taper for a group of 10 nationally ranked swimmers. There were no significant correlations, however, between performance changes and any of the other POMS subscales or the global mood disturbance scores. The authors concluded that the "addition of a psychological variable (i.e., the POMS measure of confusion) in the prediction battery is consistent with previous research suggesting that mood states are useful in monitoring training loads" (p. 1208). In much of the research, however, the subscales of vigor and fatigue have emerged as the most salient in terms of links to training load (e.g., Morgan, O'Connor, et al., 1987). Researchers have also suggested that the depression subscale might be particularly important in relation to OT (e.g., Morgan, O'Connor, et al., 1987). Thus one wonders whether the correlation Hooper et al. reported (a link between the confusion subscale and performance for a small sample of athletes) was spurious. In any case, it would be difficult to make recommendations to a coach about using the POMS results. We would not be confident in saying that elevation on any particular subscale was predictive of overtraining, indicative of poor recovery during taper, or related to performance in any way. From the results of this study, with only 10 participants, Hooper et al. suggested that one might be able to use mood disturbance to predict performance outcomes after taper. This sugges-

tion, however, was based on a significant correlation between only one subscale of the POMS and change in swimming performance, when all other subscales and the global mood score showed no significant associations to performance.

Attempting to clarify some of the disparity in the POMS and overtraining research, Martin, Andersen, and Gates (2000) examined the usefulness of the POMS for monitoring training stress in 15 cycling athletes in a well-controlled, prospective study. Martin et al. presented both group patterns and individual patterns with respect to POMS profiles and tested links between POMS scores and precise performance parameters. They found that neither the group global mood disturbance scores nor any of the subscales changed significantly in response to increases in training or following a taper period. Global mood disturbance scores tended to increase for the group following the heaviest weeks of training, but this increase was not statistically significant. With respect to individual analyses, Martin et al. reported that there were no distinct patterns involving mood profiles and performance outcomes, either during training or after taper. Among athletes who had favorable POMS profiles, some had good performance outcomes, and some had poor performance outcomes. Among athletes who had negative POMS profiles, some had good performance outcomes, and some had poor ones. Of the 15 cyclists in the study, 2 were classified as overtrained according to the criteria that performance became suppressed and remained suppressed despite apparently adequate recovery. Martin et al. pointed out that neither the global mood disturbance scores nor any of the subscale scores for these 2 cyclists appeared unique during the study. One of the *overtrainers* had some of the lowest recorded mood disturbance scores in the final weeks of the study, whereas an athlete with somewhat elevated global POMS scores had his best performance during the taper. The researchers concluded that the POMS was "not useful for differentiating whether the cycling training program represented productive overreaching or counterproductive overtraining" (p. 154). Furthermore, Martin et al. made an appeal to carefully consider the choice of dependent measures in overtraining research:

> The primary goal of training for most coaches and athletes is performance. Psychological and physiological changes during high-intensity training are, therefore, primarily of interest for monitoring

training stress when they are related in some way to performance. Without performance measures it is difficult to establish whether "mood disturbances" indicate overtraining or, instead, that overreaching is going as planned. Using predictor variables (e.g., POMS scores) as dependent variables instead of performance outcome (the real dependent variable of interest) may lead to statistically significant findings, but those findings may be of little value to coaches and athletes interested in optimizing training strategies on an individual basis. (p. 154)

Although the Martin et al. study looked at a small sample of athletes, and results might differ when larger samples of athletes are tested, the researchers have set a standard for more rigorous testing of psychological and physiological parameters in the context of overtraining and sport performance.

Naessens, Chandler, Kibler, and Driessens (2000) examined the relationships between nocturnal noradrenaline excretion measurements and POMS scores in 10 high-level soccer players. Attempting to evaluate the utility of nocturnal urinary noradrenaline (NA) excretion patterns in screening for signs of overtraining (lower NA is one *possible* sign of depression), Naessens et al. used NA measurements to predict outcomes on the total mood disturbance score of a shortened version of the POMS, as well as on the fatigue subscale alone. The investigators found that nocturnal NA excretion was moderately predictive of POMS fatigue scores. Although the researchers included a measure of performance (however imprecise—mean of weekly rated game performances by two sports journalists), they did not report any links between POMS scores and performance. Furthermore, in contradiction to the recommendation made by Martin et al. (2000), Naessens et al. used the POMS total mood disturbance and POMS fatigue subscale as dependent measures. A concern we have with this study is that the POMS subscale of fatigue appears to have been indicative more of physiological fatigue than of mood disturbance; the consistent reference to mood states as markers of overtraining, therefore, remains to be questioned.

Researchers continue to use POMS without connecting scores to actual performance (e.g., Halson, Lancaster, Jeukendrup, & Gleeson, 2003; Kenttä, Hassmén, & Raglin, 2001; Pierce, 2002). Serious issues remain unresolved in the use of the POMS for monitoring athletes' moods. The POMS was designed to monitor the moods of psychiatric outpatients, not athletes. If a person with a psychiatric condition, such as depression, begins to show elevated fatigue scores along with lowered vigor scores, then those changes might signal the onset of a new depressive episode. For athletes undergoing heavy training, such a pattern would more likely reflect that they are tired from training, not mood-disturbed or headed for depression. In many studies on athletes undergoing high-intensity training, the negative changes in global mood disturbance scores are largely accounted for by the changes in vigor and fatigue. Are these athletes really more mood-disturbed? We would say that one cannot tell from their POMS scores. Barring the presence of significant factors (outside of a heavy training load) that might influence mood, these athletes are probably more tired than mood-disturbed. Using vigor and fatigue as measures of mood and contributors to global mood disturbance with athletes in heavy training is probably one of the most serious flaws in this research.

Beyond the POMS

Broadening the focus in OT research from mood disturbance to multiple behavioral and situational measures, Kellmann, Altenburg, Lormes, and Steinacker (2001) investigated the use of the Recovery-Stress Questionnaire for Sport (RESTQ-Sport; Kellmann & Kallus, 2001) as an alternative to the POMS for evaluating the impact of athletic training. The RESTQ-Sport was constructed to measure the frequency of current stress- and recovery-associated activities from multiple perspectives (i.e., emotional, physical, and social), thus addressing a more complete picture (than does the POMS) of athletes' experiences with training and overtraining processes. Kellmann et al. looked at response patterns on the RESTQ-Sport and the POMS, along with performance changes among 54 German Junior National Team rowers over 6 weeks leading up to the Junior World Championships. As in previous POMS research, Kellmann et al. reported a dose–response pattern between RESTQ-Sport scores and training loads. Also as in previous POMS research, however, Kellmann et al. did not test or report statistical links between RESTQ-Sport scores and performance. Nevertheless, the research on the RESTQ-Sport is relatively new, and it may be just a matter of time before the RESTQ-Sport is tested rigorously in association with meaningful performance outcomes. Kellmann et al. pointed out the advantages of the RESTQ-Sport in applied settings in terms of providing much more information than the POMS does about athletes' behaviors, stressors, and recovery states:

> For the POMS, we have the "iceberg profile," which primarily consists of negative mood states and only one aspect dealing with positive states of mood, while the RESTQ-Sport gives us a detailed picture of the athletes' state. Concrete solutions to current problems can be derived from the up-to-date recovery-stress profile and this profile might obviously be used to derive specific intervention strategies. . . . The RESTQ-Sport provides coaches, sport psychologists, and athletes with important information during the process of training. (pp. 163–164)

Although the usefulness of the RESTQ-Sport for predicting performance has yet to be established, it seems that it could be a practical tool at the individual level to provide important parties with information about how athletes may or may not be coping with a training program. The POMS has not proven to be a reliable tool for predicting how athletes will perform after a cycle of training and recovery, though it may still have some clinical utility for opening a discussion between, for example, a sport psychologist and an athlete with a significantly elevated mood disturbance score. The RESTQ-Sport might be an important step toward developing a more holistic picture of an athlete's psychosocial and physical responses to training and recovery. Nevertheless, predictive utility of RESTQ-Sport for performance outcomes remain to be established.

Comments on the Elusiveness of Overtraining Markers

With respect to markers and diagnosis of OT, researchers have noted that none of the hypothetical markers for OT syndrome is unequivocal (Armstrong & VanHeest, 2002), and a recent "critical review of existing scientific literature leads to the disappointing conclusion that the tools available for overtraining syndrome diagnosis have not improved much in the last years of overtraining research" (Urhausen & Kindermann, 2002, p. 100). Although the physiological assessment research is an invaluable contribution to the body of knowledge on OT, some of the inconclusive research findings may have resulted from adherence to a dose–response model of training and recovery (Morton, 1997). From this perspective,

it is assumed that, if the body is pushed too hard with high workloads (*dose*), physiological and psychological disruptions will occur (*response*), from which an athlete needs time to recover. The dose–response model, however, puts the focus mostly on what is happening with the training load (or dose), excluding some issues relevant to OT, such as individual differences in stress tolerance, existence of multiple stressors beyond the training program, and potential causal mechanisms driving the maladaptive OT behaviors. Furthermore, the responses being measured (e.g., physiological or psychological markers, performance decrements) have often turned out to be inaccurate indicators of OT. Although it would be useful for coaches, athletes, and sport scientists to have an established selection of specific markers that would indicate when an athlete is beginning to overtrain, the processes underlying athletic training, recovery, and OT, appear to be too complex to be measured easily. Nonetheless, researchers have recently moved toward a more holistic view of the training process by looking at the balance between stress and recovery (Kellmann, 2002; Kenttä & Hassmén, 2002), emphasizing the recovery aspects of training, and assessing many aspects of athletes' lives both inside and outside of physical training.

With respect to the psychological assessment research, some researchers have argued that psychological testing is the most effective in detecting OT at an early stage (Kellmann & Günther, 2000; O'Connor, 1998; Shephard & Sheck, 1994). Nonetheless, as with physiological research, the results of psychological assessment (especially POMS) research have proven equivocal because many of the markers have also been associated with intense training that did not lead to performance decrements (Rowbottom, Keast, & Morton, 1998). Steinacker and Lehmann (2002) stated that "performance is the most important parameter for monitoring training adaptation. Maximum performance during a standardized test is therefore the gold standard for evaluating exercise capacity and monitoring training" (p. 107). Along these lines of emphasizing performance, Kaplan (1990) put forward an important message about behavioral outcomes in health care research and delivery that is relevant to OT contexts. He pointed to behavior as the central outcome for health care, asking whether interventions have helped patients improve in practical aspects of their lives, such as walking, talking, or caring for themselves. In the OT context, Martin et al. (2000) commented on Kaplan's emphasis on behavioral outcomes:

This same focus on behavioral outcomes should apply to overtraining research. Changes in POMS scores and physiological variables are only of interest to the coach and the athlete if they relate to performance. If POMS scores or blood chemistry (e.g., cortisol levels) change significantly over the course of an intense training program and taper, they are of little practical interest to coaches and athletes unless they are related to the outcome of interest, performance. (p. 142)

Some researchers (Martin et al., 2000; Rowbottom et al., 1998) have suggested that methodological flaws, such as conducting overload training research without a taper and not measuring performance and OT markers after a taper, may mean that many supposed OT markers are no more than indicators of current training load. Without knowing what happens to performance after taper, one cannot say anything more than that the assessment markers are good indicators of intense training (a negative POMS profile or skewed physiological assays associated with intense training loads before taper could be associated with either performance improvement or decrement after taper, depending on the individual).

Psychological and physiological assessment might continue to prove useful in applied settings to inform athletes and coaches when *something* is going on in athletes' lives, in terms of stress or recovery responses to training. Future research, however, could be directed at uncovering causal antecedents and risk factors for OT. Moving away from the dose–response approach, and with the emphasis on stress–recovery balance as described in their conceptual model, Kenttä and Hassmén (2002) have helped initiate the move to a more holistic understanding of athletes' experiences with OT. Several researchers (Kellmann, 2002; Kenttä & Hassmén, 1998; Meyers & Whelan, 1998; Raglin, 1993) have emphasized the individual differences in athletes' responses to training and life stressors. Thus, further research into the meanings, experiences, and causes of OT in individual athletes could prove fruitful.

OTHER OVERTRAINING OUTCOMES

With regard to the variations in outcomes (e.g., staleness, burnout, overtraining syndrome), it may be important to consider injury and illness along with OT syndrome as possible consequences of OT. Although the connections of OT with injury and illness have been mentioned in the literature (e.g., Fry et al. 1991; Kuipers, 1996), an in-depth exploration of the connections is not available. In taking a broader view of OT, we see overuse injuries as an OT result stemming from many of the same factors that lead to classic OT syndrome (e.g., perfectionism, psychosocial stressors, intrapsychic needs, poor coaching, sport cultural traditions).

OT and Injury

One reason there are not many studies connecting OT with injury directly is that the injuries usually stem from overuse and are not seen in a broader context of athletes overdoing training. Many of these injuries probably occur well before OT syndrome kicks in or occur in sports where OT syndrome would be rare (e.g., golf). So the connection of overuse injuries with behaviors that lead to overtraining syndrome has not received much attention. The literature is seriously limited concerning this connection. As O'Toole (1998) suggested,

> Increased susceptibility to musculoskeletal injury or infections, such as head colds, may be indicators of a state of overreaching or overtraining, but may be misinterpreted as isolated, local problems rather than manifestations of the overtraining syndrome. (p. 13)

The literature on overuse injuries, however, is vast. We would suggest that many overuse injuries result from athletes engaging in overtraining behaviors that stem from the environmental pressures to perform and from intrapersonal pressures to strive for excellence. Flynn (1998) did suggest that, in some sports, OT might lead to musculoskeletal breakdown before the onset of OT syndrome. For example, excessive training volumes in swimming might lead to OT syndrome, whereas running at comparable volumes would more likely lead to joint or other musculoskeletal injury due to the repetitive impact activity of the sport. Despite differences among sports, it seems possible that overtraining, as defined by an imbalance between stressors and recovery, might lead to injuries in any sport. Because of the severe limitations in the literature connecting overtraining behaviors with injury, we will have to rely primarily on one study (Kibler & Chandler, 1998) for suggestions about overtraining injury outcomes. Kibler and Chandler have discussed the potential interaction between OT and musculoskeletal breakdown as follows:

> Inappropriate volume or intensity of exercise may cause maladaptive cellular or tissue responses due to an imbalance between load and recovery. These maladaptive responses occur to some extent in most all sports, but they can certainly become part of the overtraining syndrome. The maladaptive responses may be objectively documented as distinct musculoskeletal injuries, such as alterations in muscle strength, flexibility, or balance, changes in joint range of motion, or stress reactions in bone. (p. 169)

Although Kibler and Chandler (1998) suggested that the precise mechanisms of musculoskeletal overtraining are not comprehensively understood, they pointed out that the maladaptations seem to originate from disruptions in cellular homeostasis. Furthermore, Kibler and Chandler noted that although "cellular disruptions occur in all athletes, the overtrained athlete is particularly susceptible to maladaptation and injury as the result of chronic overloads and disruptions" (p. 170). From this description of musculoskeletal breakdown in the context of overtraining, one might infer that athletes with the highest training loads and intensities are most likely to be at risk for OT and its outcomes. Nonetheless, Kibler and Chandler were careful to clarify that the training load is important only as a "relative load compared to the muscle's ability to protect itself against strain. Normal loads on weakened muscles, a relative force overload, are as capable of causing strain as supernormal loads on normal muscles, an absolute force overload" (p. 171).

In reviewing Kibler and Chandler's (1998) discussions of musculoskeletal adaptations to training overloads, it seems that the relationship between overtraining and injury is subtle and sometimes hard to detect. According to these authors, adaptations associated with overload training may not appear as overt clinical symptoms but may manifest in the system as mechanical alterations or decreases in performance efficiency. Furthermore, injuries from overload often arise from an accumulation of stress, with a gradual onset, making detection difficult and increasing the likelihood of misattributions about the causes of injuries. Kibler and Chandler pointed out how overtraining in the musculoskeletal system can lead to injury, a process they described as a "cascade-to-overload injury":

> Tissue compromise can more frequently create functional biomechanical deficits as a result of alterations in flexibility, strength, strength balance, or skeletal reaction. The athlete attempts to compensate for these deficits by adopting alternate patterns of movement, position, and activity. These patterns are usually less efficient, creating even more overload, thereby closing the circle. (pp. 174–175)

If athletes begin a training cycle with muscle weaknesses, muscle imbalances, or areas of inflexibility, they may be predisposed to overload-related injuries. Kibler and Chandler (1998) noted that such athletes would be susceptible to OT when exposed to the extrinsic demands of their sports and might experience tissue failure and clinical symptoms if they continued to train in this susceptible state. In particular, Kibler and Chandler remarked that stress fractures may be manifestations of OT, given that the fractures often occur because the skeletal system cannot keep up with the demands of overload training. In describing the potential mechanisms of skeletal injury, Kibler and Chandler suggested that weakened or fatigued muscles may be unable to handle either absolute or relative force overloads, thus transferring those forces to the skeletal system and producing stress reactions. In relation to soft tissue injuries, the authors also noted that "inappropriate overload can be a causal factor in the acute muscle strain both in the form of abnormal biomechanics and a decrease in the ability of the muscle to protect itself" (p. 178). Timing also seems to be a crucial factor in the etiology of OT injuries. Oftentimes, athletes returning from previous injuries begin intense training too soon, thus risking further overloads and chronic OT injury problems. It seems that coaches and athletes may not always be aware of the myriad ways that stress–recovery imbalances can occur, and may not look at OT as playing a role in the lead-up to injury. Kibler and Chandler noted that OT might be overlooked in analyzing common sport injuries:

> Even in a situation where a hamstring injury is seemingly unrelated to overtraining, hamstring tightness from continued use or a muscle strength imbalance may be a hidden contributor to the injury. . . . Sub-clinical soft tissue injuries are injuries that may not be recognized as injuries by the athlete or the coach and are often overlooked as a possible causative factor of more severe injuries. (p. 179)

In reviewing the OT literature, the interactions between injury and OT are often not emphasized, yet it appears that injury can be both a possible outcome of and a contributing factor to OT. In highlighting some of the potential injury–OT interactions, Kibler and Chandler (1998) noted that "musculoskeletal maladaptations and injuries can be a warning signal to the athlete and coach that the volume or intensity of training is too high, and overtraining is a possible causative factor" in injury outcome (p. 186).

OT and Illness

OT has also been associated with illnesses such as head colds, allergic reactions, and upper respiratory tract infections (Armstrong & VanHeest, 2002; Costill, 1986; Jokl, 1974; Mackinnon & Hooper, 1994; Nieman, 1998; Steinacker & Lehmann, 2002; Weinstein, 1973). In contrast to findings that regular physical activity has positive effects on immune system function, there is evidence that high training loads will increase the risk of infections (Nieman & Pedersen, 1999; Steinacker & Lehmann, 2002).

If athletes begin a training cycle with muscle weaknesses, muscle imbalances, or areas of inflexibility, they may be predisposed to overload-related injuries.

It appears that prolonged, exhausting exercise taxes the immune system and may result in clinically significant alterations in immune function. The literature on exercise, immunology, and illness is vast (see Nieman & Pedersen, 1999, for a review) and beyond the scope of this chapter. We will focus briefly on heavy workloads and the risk of upper respiratory tract infections (URTIs; Gleeson et al., 2000; Nieman, 2000).

According to Nieman (1997), "the epidemiological data suggest that endurance athletes are at increased risk for upper respiratory tract infections (URTI) during periods of heavy training and the 1–2 week period after marathon-type race events" (p. 344). Later, Nieman (1998) gave the same warning for the "1–2 week period following prolonged and intensive aerobic exercise" (p. 193). This 1- to 2-week period of risk may derive from the "open window" of 3 to 72 hours of acute immune system changes following heavy bouts of exercise, changes that may allow for infections to take hold, incubate, and manifest in the next week

or so (Nieman, 2000). This open-window hypothesis, although intuitively attractive, has not been studied extensively yet and needs further research. Similarly, in reviewing the literature on illness in sport and exercise, Weidner (1994) showed that URTIs are the most common infection among elite athletes. There is evidence for a J-shaped model for exercise workload plotted against URTI rates. Sedentary people have about the average amount of URTIs each year. People who exercise moderately have below average rates, and exercisers with very high training loads have above average yearly rates (Nieman, 1997).

Looking more closely at illness–OT interactions, Mackinnon (1998) posed the question, "Does illness due to intensive training cause or contribute to overtraining?" (p. 234). Mackinnon stated that frequent illnesses are considered common outcomes or symptoms of OT, and noted similarities between symptoms of OT syndrome and those of infectious illness, such as persistent fatigue, decreased performance, inability to train effectively, muscle soreness, and lethargy. Nonetheless, Mackinnon suggested that "the presence or absence of infectious illness should be discounted [i.e., actively ruled out] or documented when diagnosing OT among athletes" (p. 234). Mackinnon did not seem to answer the question of whether illness contributes to OT syndrome. Rather, she seemed to say that the symptoms of illness are similar to those of OT syndrome, and went on to comment that the presence or absence of illness should be part of assessing OT.

In conclusion, it seems that illness might be both a contributor to and an outcome of OT. When athletes are in a state of stress–recovery imbalance, they become more susceptible to infections and illnesses, which further stress their bodies, leading to higher risk for greater imbalances and OT syndrome. The initial stress–recovery imbalance is part of the OT process, and the interaction of this OT process with illness is circular; each one may contribute to and be an outcome of the other. Gotovtseva, Surkina, and Uchakin (1998) made a clear statement that illness is part of the OT etiology: "Along with the other classic symptoms of overreaching and overtraining, immune dysfunction and frequent colds have been found in overreached and overtrained athletes, and thus may be considered as markers of this athletic pathology" (p. 265).

CONCLUSIONS

In this chapter, we have considered research addressing a number of basic questions about overtraining. In prevalence research, sport scientists have examined the scale of overtraining problems, especially in elite sport. We addressed research that has explored physiological and psychological markers of overtraining. Researchers have also focused on understanding the development of overtraining and on investigating processes by which athletes overtrain, as distinct from the outcome of these processes—that is, overtraining syndrome. Based on our argument that outcomes of the processes of overtraining and underrecovery include injury and illness, as well as overtraining syndrome, we also discussed research on processes involved in injury and illness outcomes.

In general, our conclusions are that, although a substantial amount of research effort has been focused on overtraining, we have few definitive answers. Claims are often based on research with limited sample sizes. Sometimes positive findings come from studies in which researchers monitored multiple measures and only one or two measures produced significant results. Often, researchers have not replicated noteworthy findings in other studies with similar designs or measures, and researchers asking similar questions have employed differing measures, making direct comparison of results difficult.

We acknowledge that, for various reasons, many types of research on overtraining are not easy to conduct. Experimental studies aimed at manipulating training loads and trying to produce overtraining syndrome have obvious ethical problems, and we would not want to see such research conducted. Another major issue concerns sampling. Answering questions about prevalence requires large samples, but researchers cannot conduct the kind of large-scale, cross-sectional research that identifies the current prevalence of diabetes or asthma, because it is not known when somebody is going to overtrain. To address this problem, researchers have often adopted retrospective approaches to examine the prevalence of overtraining, but this tack involves having athletes make subjective judgments about whether they have overtrained during some period in the past, such as the last 12 months. Investigators must view with caution the reliability of such subjective judgments based on memories of past events, and research on identification of overtraining based on precise scientific markers is inconclusive.

In a similar way, the weaknesses we have identified in the research on each aspect of the overtraining phenomenon reviewed in this chapter often influence the evaluation of research on other characteristics of overtraining. For example, we suggest that doubt about whether the conventional markers of overtraining really are valid indicators reflects on their potential for meaningful use in research into processes and outcomes. Similarly, we propose that the potential for overtraining outcomes (e.g., illness or injury) in elite athletes is probably underestimated, because athletes often get injured or ill before overtraining syndrome is identified. We base this proposition on research indicating that the processes leading to injury and illness among elite athletes have much in common with overtraining processes. We argue that an acute injury or debilitating illness is usually readily identifiable. Upon identification of injury or illness, athletes who are in the process of overtraining are forced to stop training, inadvertently managing their overtraining, at least temporarily.

Despite our concerns about research on the prevalence, processes, and markers of overtraining, we recognize that overtraining is a serious problem that affects many athletes. This kind of research, aimed at strengthening our understanding of overtraining phenomena, must continue and be refined. We propose that it is important to focus research on early identification of overtraining and underrecovery, so that sport science practitioners can intervene to minimize the risk of athletes becoming damaged. We turn our attention to this issue in the following chapter.

CHAPTER 3

WHAT BRINGS IT ON?
Risk Factors for Overtraining

In chapter 2, we showed how quantitative research has largely supported (with some reservations) the contention that overtraining (OT) and associated phenomena, such as injury and illness, are common in sport, especially at the elite level where physical and psychological pressures can be substantial. The picture that emerged from chapter 2 reflects current understanding in the field of the extent of overtraining in high-level sport, as well as those variables that appear to be associated with the overtraining process and its outcomes of overtraining syndrome, injury, and illness. For most part, what that research has told us is of only limited value to athletes, coaches, and sport science practitioners, whose aim is to prevent people from overtraining. Unfortunately, by the time we can see the physiological, psychological, and behavioral signs of overtraining, it is usually too late to stop it; the athlete has overtrained, or has become sick or injured. To intervene before athletes have overtrained, we need to know what factors lead to overtraining: What triggers the overtraining process? What are the risk factors?

Researchers have rarely examined the risk factors for overtraining directly. A range of research, however, does provide clues about variables that might influence overtraining. That research primarily comprises qualitative studies of two kinds. Some suggestions have emerged from studies of the views of experts who work with athletes, including coaches, trainers, sport scientists, and sports medicine personnel. Researchers have also identified potential risk factors in investigations of athletes,

focusing on aspects of their thoughts, feelings, and behaviors associated with times of overtraining.

If we can identify the risk factors, as well as the signs and symptoms associated with those factors, then we may be able to construct an "early warning system" that alerts coaches, athletes, and support staff to potential problems. In medicine, the earlier a disease process is detected, the better the prognosis will usually be, assuming, of course, that the disease is treatable. Because there are so many variables (risk factors) involved in overtraining, and because each athlete responds to training differently, and probably has different constellations of contributing factors, there will not be a one-treatment-fits-all solution. Still, overtraining is a process that has the potential to be 100% treatable.

Well before we can get to a solution or intervention phase, however, we have to know what those risk factors are, and that is the focus of this chapter. First, we discuss what we mean by risk factors, then we look at research focused on the views of experts who have suggested risk factors. Next, we address research exploring the experiences of athletes in various contexts, such as preparation for and performance in major events, where athletes sometimes report on experiences that appear to lead to overtraining. To conclude the chapter, we pull together the descriptive material discussed here and in chapters 1 and 2 to present the rationale for the in-depth studies we have conducted on experts' and athletes' perspectives on risk factors for overtraining.

Because there are so many risk factors, and because each athlete responds to training differently, there will not be a one-treatment-fits-all solution. Still, it is a process that can be 100% treatable.

RISK FACTORS FOR OVERTRAINING

When we talk about risk factors, we are usually referring to variables that influence the occurrence of negative or undesirable behaviors or outcomes. For example, based on research, it is well established that smoking and depression are independent risk factors for heart disease. Thus, people who smoke and people who are depressed are more likely to have heart attacks than those who don't smoke or who are not depressed. For some time, organizations committed to reducing the incidence of heart disease, particularly death from heart attacks, have implemented interventions to reduce smoking in many countries around the world. The role of depression in heart disease has been identified only recently, so strategies to manage depression have yet to be implemented by organizations fighting heart disease. A similar example in the field of sport psychology relates to the incidence of injury. Researchers have demonstrated that psychosocial stress is a major risk factor for injury in sport and in dance (Noh, Morris, & Andersen, 2005; Williams & Andersen, 1998). Thus, athletes or dancers who experience high levels of stress but lack the resources to cope effectively are more likely to get injured. Now that this connection is established, researchers are testing interventions to help athletes and dancers manage their stress more effectively. To reduce the occurrence of overtraining, researchers and practitioners in sport and other intensive performance contexts need to identify those variables that, when present, or in some cases absent, increase the likelihood that overtraining will occur. Those variables are the risk factors. Only after those factors are identified can researchers develop and test interventions for practitioners to use in helping athletes minimize the risks that they will overtrain.

Although qualitative studies have not been conducted specifically on OT risk factors, interviews with and case studies and observations of athletes by researchers have provided clues about what the OT experience is like for athletes, and what personal and situational variables add to the stress loads that put them at risk for OT, illness, and injury. In the following sections, we outline perspectives on the personal and situational variables affecting athletes' training and recovery. This research comes from both experts (mostly researchers and coaches) and athletes.

Experts' Perspectives on Risk Factors for Overtraining

Researchers in a large number of OT studies and review articles have made anecdotal comments about the potential risk factors for overtraining. In some cases, these researchers have presented interview data that suggest potential OT risk factors. Krane, Greenleaf, and Snow (1997) reported that an elite gymnast who overtrained was characterized by disordered eating, high levels of ego-involved goal orientation, perfectionism, a win-at-all-costs attitude, maladaptive responses to failure (e.g., increasing an already excessive training load), and an ability to rationalize excessive training practices. The gymnast also was surrounded by parents and coaches who always pressured her to win and supported her rationalizing and other maladaptive behaviors. In an interview study, Gould, Tuffey, Udry, and Loehr (1997) identified overtraining as a major contributing factor to burnout in junior tennis players; one player who overtrained was characterized by a high level of perfectionism and unrealistic expectations, and was subjected to elevated parental criticism, expectations, and emphasis on winning. This athlete believed that increasing her training load (when it was already very heavy) was the only route to success. The researchers also discovered that a second tennis player who overtrained did not report such high levels of perfectionism or parental pressure as other players, but displayed a profile suggesting "supermotivation" combined with unrealistic goals. This player held the belief that hard work, rather than talent, would bring him success. Finally, in an interview study of Olympic athletes, Gould, Guinan, Greenleaf, Medbery, and Peterson (1999) revealed that among many athletes who identified overtraining as a major contributor to their failures, lack of good coach–athlete communication and poor timing of selection processes (e.g., Olympic trials for team selection held many months before the games) were cited most commonly as damaging factors. Other researchers, noting similar qualities in athletes they observed, have commented on circumstances that may put athletes at risk for overtraining such as lack of recovery time (e.g., Kenttä & Hassmén, 2002).

Lack of Recovery Time

Fry, Morton, and Keast (1991) stated that "the most outstanding training-related factor leading to overtraining is reported to be the failure to include enough regeneration units [recovery time] in the training programme" (pp. 39–40). There may, however, be many factors that affect recovery. For some athletes, the training stimulus may be the most important factor affecting recovery, if the workload is of a high volume and intensity. For others, factors outside of training may impede adequate recovery, such as work, school, or family commitments. Lehmann, Foster, and Keul (1993) suggested that it is important to look at each athlete individually, and Kellmann noted (2002) that "inter-individual differences in recovery potential, exercise capacity, non-training stressors, and stress tolerance may explain the different degrees of vulnerability experienced by athletes under identical training conditions" (p. 16). There also may be subtle influences on athletes' attitudes toward and behaviors surrounding recovery: input from coaches and parents; pressure from sport culture; and attitudes of peers, training partners, and other athletes. Brustad and Ritter-Taylor (1997) noted that coaches and others frequently endorse attitudes such as "no pain, no gain" and "more is always better," which create cultures of risk instead of promoting positive self-care and self-awareness. Kenttä and Hassmén (2002) noted that sudden increases in nontraining stressors add to the total stress, which can reach a level where a person experiences a lack of recovery. Therefore, overtraining can occur even during moderate levels of physical training if there are coexistent high levels of psychosocial stress.

In a description similar to Fry et al.'s (1991) comments about lack of recovery, Hanin (2002) discussed risk factors for OT in terms of barriers to effective recovery and rest. He suggested that athletes and coaches may underestimate the importance of systematically matching workload with adequate rest. He pointed out that this underestimation may be reinforced by the values held by some sport cultures, subcultures, and athletes, wherein quantity (involving intensity and volume) is emphasized over quality. Hanin also observed that athletes' responses to their own performances can affect how they balance training and recovery, potentially motivating them to push excessively in training:

> In the case of poor performance (underperformance due to fatigue or problems with technique), an athlete continues to work intensively to eliminate uncertainty and to enhance self-confidence. However, athletes usually are unable to break this vicious circle and even do not dare to take

a good break and correct this situation. In the case of successful (better-than-expected) performance, an athlete can be so over-excited with positive emotions that he does not notice the signs of fatigue and . . . continues to do excessive work until it is too late. (p. 210)

Guilt About Not Working Hard Enough

Botterill and Wilson (2002) observed that "guilt about not working hard enough and being intense all the time" (p. 144) appear to be important risk factors for overtraining and can impede recovery and rehabilitation. Circumstances such as the lead-up to a big competition, or the time immediately after it, may also increase the risk of overtraining, possibly due to the addition of heightened mental and emotional demands to the already existing physical demands of preparing for and performing in competitions. Botterill and Wilson also recognized the potential harm of emotional buildup, commenting that "repressed, denied, or unprocessed emotions" (p. 150) can be sources of conscious and unconscious conflict and stress.

Change in Environment

Kellmann (2002) noted that sometimes a simple change in environment can have a profound effect on an athlete's recovery:

> If an athlete's accommodation is close to a loud street, her rest may be disturbed day and night. However, if an athlete is used to living in a loud neighborhood, she might have no problem sleeping through loud noises, but instead may get irritated by an absolutely quiet environment. (p. 9)

Disruption of Sleep

Kellmann (2002) also suggested that a disruption of sleep can be a direct result of emotional disturbances in an athlete's life, such as family or relationship conflicts. Kellmann recognized that although athletes can compensate for a lack of sleep or other recovery activities in the short term, they will eventually risk developing an overtraining syndrome.

Competitive Environments

In many sports, competitive environments and intrateam rivalries can lead to disruptions in recovery, exemplified in the following anecdote from Kellmann (2002):

The coach of the Canadian male speed skating team planned a training schedule that included a day off as the key element for recovery purposes. The coach did not tell the athletes what to do for recovery, so they decided to go for a bike ride in the mountains. The purpose of the bike ride was to relax, be with the team, and get refreshed by the scenery of the Canadian Rocky Mountains. However, the athletes soon turned the relaxing bike ride into a competition that left no room for physiological recovery at all. . . . Since the athletes knew that the coach would not appreciate their bike ride competition, they did not tell him, and the next day practice continued based on the regular schedule. The next physiological stressor was set, and some days later the coach was surprised by the performance decline. (p. 9)

Kellmann (2002) observed that, although the athletes in this story may have experienced some social recovery from the fun ride, their competitiveness resulted in a physiological stressor that could have disrupted the program prescribed by the coach. The coach appeared to be unaware of the risk for OT. In other cases, however, coaches might even acknowledge that they are risking OT with aggressive approaches to training, illustrated by the following quote from Gould and Dieffenbach (2002):

> If five is good then 50 is better. . . . The year before the Olympics, they [my athletes] would do sets of 50 jumpies and this year they are doing sets of 700 jumpies. If I can push them more and more, when they finally get there, they will be great. However, this pushing can blow up in your face—like you finally get to the Olympics and are exhausted. (p. 26)

Maladaptive Responses to Poor Performance

Hawley and Schoene (2003) noted that athletes may display maladaptive responses to poor performance, where frustration leads them to train harder in response to plateaus or declines. The problem with increasing training efforts, in cases where stress loads are already high, is that it increases fatigue and results in further decrements in performance, thus initiating a vicious cycle of heavy training, chronic fatigue, poor performance, and frustration.

In summary, these various recent observations made by researchers indicate that understanding what puts athletes at risk for OT may require looking into many aspects of athletes' lives, both personal and situational. Although there have been no systematic investigations into OT risk factors, it is possible to summarize the observations made by researchers and experts thus far. Table 3.1 illustrates many of the variables that may arise in relation to risk for OT. Some of the factors seem readily apparent, such as high-volume or high-intensity training and inadequate recovery. Others are less obvious (e.g., an athlete being at risk when at a physiological peak), thus highlighting the importance of gathering as much information as possible about athletes when assessing the risk for OT.

Table 3.1   OT Risk Factors Identified in the Literature by Researchers

Risk Factor	Sources
Training issues (program & schedule)	
High-volume, high-intensity training	Brown et al. (1983); Budgett (1998); Derman et al. (1997); Foster (1998); Hollander & Meyers (1995); Kenttä & Hassmén (2002); Kuipers (1996); Kuipers & Keizer (1988); Lehmann et al. (1993); Uusitalo (2001); Wallace (1998)
High training monotony; lack of periodization	Armstrong & VanHeest (2002); Budgett (1998); Foster (1998); Hollander & Meyers (1995); Wallace (1998)
Failure to include recovery in training program; lack of rest days	Kenttä & Hassmén (2002); Kuipers (1996); Kuipers & Keizer (1988); Wallace (1998)
Sudden increases in training load or intensity (particularly lactate training, and especially following breaks due to injury or illness)	Brown et al. (1983); Budgett (1998); Hooper & Mackinnon (1995); Kuipers & Keizer (1988)
Lack of seasonal layoffs	Hooper & Mackinnon (1995)
High volume of dry-land training or cross-training	Hooper et al. (1995)
Frequent competition or year-round competition	Brown et al. (1983); Derman et al. (1997); Kuipers (1996); Kuipers & Keizer (1988); Wallace (1998)
Transitions in training programs—usually from winter low-intensity to spring interval and higher-intensity programs	Budgett (1998)
Time of season—especially just prior to and during competition; competition and selection	Budgett (1998); Hawley & Schoene (2003); Uusitalo (2001)
New training environment	McCann (1995)
Lack of training program flexibility and individualization: team sports where coaches do not have leeway to take individual training tolerance into consideration when planning practice; individual sports with one training program for all athletes	Hooper & Mackinnon (1995); Levin (1991)
Lack of proper taper	Gould et al. (2002); Levin (1991)

(continued)

Table 3.1 OT Risk Factors Identified in the Literature by Researchers, *continued*

Risk Factor	Sources
Lack of objectivity when athlete is doing own training, training without coach or partner, or training with significantly more skilled or more physically fit athletes	Brown et al. (1983)
Lack of monitoring for signs of overtraining	Hooper & Mackinnon (1995); Committee on Sports Medicine and Fitness (2000)

Situational and environmental stressors

Travel (especially across time zones); jet lag	Derman et al. (1997); Gould et al. (2002); Kuipers (1996); Uusitalo (2001); Wallace (1998)
Changes in training environment, altitude, temperature, humidity	Armstrong & VanHeest (2002); Kuipers & Keizer (1988); Uusitalo (2001); Wallace (1998)
Moving house or other economic stressors	Beil (1988); Uusitalo (2001)
New national team status	McCann (1995)
Increases in employment workload and other occupational stressors	Armstrong & VanHeest (2002); Kenttä & Hassmén (2002); Kuipers & Keizer (1988); Lehmann et al. (1993)
Poor performance at competition	Hawley & Schoene (2003)
Problems and obligations in school; increases in academic workload	Armstrong & VanHeest (2002); Kuipers & Keizer (1988); Kellmann (2002); Kenttä & Hassmén (2002); Lehmann et al. (1993)
Sport specialization at an early age	Committee on Sports Medicine and Fitness (2000)
Participating at too high a level for ability (especially among youth athletes)	Committee on Sports Medicine and Fitness (2000)

People issues (coaches, parents, others)

Conflicts with coaches; relationship problems with friends, teammates, staff, or parents	Armstrong & VanHeest (2002); Hollander & Meyers (1995); Kenttä & Hassmén (2002); Kuipers & Keizer (1988); Wallace (1998)
Excessive expectations or unrealistic goals from a coach or family	Armstrong & VanHeest (2002); Hollander & Meyers (1995); Kuipers & Keizer (1988)
Emotional stress from major life events (e.g., marriage, new baby in family, illness, conflicts with partners, parents' divorce)	Kellmann (2002); Hollander & Meyers (1995); Kenttä & Hassmén (2002)

Athletes' physical issues

Premature return from injury	Budgett (1998)
Physical illness, allergies, disease, or infections	Kuipers (1996); Kuipers & Keizer (1988); Raglin (1993); Uusitalo (2001); Wallace (1998)
Poor or inadequate sleep	Derman et al. (1997); Kenttä & Hassmén (2002); Uusitalo (2001)

Risk Factor	Sources
Poor or inadequate nutrition; possibly inadequate caloric intake (especially carbohydrates); potential nutrient, vitamin, or mineral deficiency; iron deficiency; dehydration	Committee on Sports Medicine and Fitness (2000); Derman et al. (1997); Hollander & Meyers (1995); Hooper et al. (1995); Kenttä & Hassmén (2002); Kuipers (1996); Kuipers & Keizer (1988); Uusitalo (2001); Wallace (1998)
Adolescent athletes during growth spurts; overloading developing bodies	Beil (1988); Committee on Sports Medicine and Fitness (2000)
Athletes with a substantial injury history or experiences with overtraining	Hollander & Meyers (1995); Raglin (1993)
Prolonged amenorrhea in female athletes leading to diminished bone mass	Beil (1988); Committee on Sports Medicine and Fitness (2000)
Low tolerance for physical or psychological stress loads (predisposition); poor recovery potential	Kenttä & Hassmén (2002); Lehmann et al. (1993); Uusitalo (2001)
Athletes at their physiological peaks on the threshold of overtraining	Armstrong & VanHeest (2002)

Athletes' beliefs, behaviors, and attitudes

Success—rapid rise to elite level (especially for young athletes); new personal bests causing athletes to believe training harder will bring even greater success	Budgett (1998); McCann (1995)
Unrealistic role models—athletes comparing themselves with and trying to keep up with faster, better-skilled athletes; comparing with successful others who train at high volumes, beyond the athlete's current capacity	Brown et al. (1983), Budgett (1998)
Desperation in response to mediocre performance	Budgett (1998)
Very high levels of motivation to achieve success; motivation to set a new standard (e.g., world record)	Budgett (1998); Hollander & Meyers (1995); Kuipers & Keizer (1988); Levin (1991)
Maladaptive responses to underperformance (e.g., increasing training load or not decreasing other stressors when loads are already high)	Foster (1998); Kenttä & Hassmén (2002); Raglin (1993)
Belief that feeling fatigued is equivalent to being unfit, requiring increases in training (when training loads are already high)	Levin (1991)
Unrealistic goals set by athlete	Hollander & Meyers (1995)
Fear of failure	Kuipers & Keizer (1988)
Personality structure—ongoing personal or emotional problems	Hollander & Meyers (1995)
Difficulties with time management (vis-à-vis practice, school, friends)	Kellmann 2002
Fear of being undertrained—more-is-better philosophy	Brown et al. (1983); McCann (1995)

Athletes' Perspectives on Risk Factors for Overtraining

Many athletes have spoken about their experiences with stressors they confronted during training and recovery. In several different studies and reviews related to OT, researchers have presented direct quotations from athletes, illustrating their experiences with high-pressure situations, pushy coaches, and overinvolved parents, as well as athletes' own attitudes toward training and recovery.

External Stressors

External stressors are extrapersonal variables that may influence athletes' stress–recovery balance. These stressors include other people–such as coaches, parents, friends, and other family members–as well as situations and environments. Training venues can be cold and drafty. Atmospheric conditions, such as smog, can cause physical and physiological stress (e.g., physical irritation to eyes and lungs, physiological responses such as an asthma attack). Workplace tension can drain athletes who need to be employed. The number of potential external stressors is nearly endless; we will start with significant other people.

Coaches Pushing Too Hard

Coaches have a significant influence on how hard athletes push in training, how they pursue and experience recovery activities, and how they feel about themselves, emotionally and physically. Some athletes are aware that their coaches may push too hard, risking adverse outcomes, as illustrated by an athlete quoted in Gould and Dieffenbach (2002): "My coach is a real pusher, to the point where I think he pushes too hard. I think I would be better if I did not train as hard" (p. 25). For an in-depth example of a coach who pushed too hard, see chapter 9.

Coaches Not Accounting for Individual Differences

Coaches sometimes do not account for athletes' individual differences in training and recovery capacities:

> I think the coach failed to see the individual needs of players. Some people just couldn't practice for three hours in 90-degree heat. It got to them. Quite a few were sick, off and on, and half our team was injured. (Wrisberg & Johnson, 2002, p. 264)

Abusive Coaching

In other cases, athletes have reported that their coaches were simply abusive:

> All [the coach] knew how to do was bitch at us. She made us feel like we were fat . . . real big. She called me names and told me how mentally disabled I was. She had something for everybody—I just happened to be the retarded one in her eyes. She liked to make cracks about our bodies. We were already pretty self-conscious about being big. So around her, we always felt so fat— just horrible and ugly. And the uniforms didn't help us one bit because they were real short. . . . We just never felt very good about ourselves and she had a lot to do with that. (Wrisberg & Johnson, 2002, p. 264)

Such an emotionally charged environment can prompt some athletes to go to extremes in training, eating behaviors, or other potentially harmful activities. Other forms of abusive behavior by coaches include pressuring athletes to perform when they might be unfit to do so. Wrisberg and Johnson (2002) noted that, as an outcome of OT, injury poses a serious threat to the well-being of athletes in the long term, especially when coaches mishandle injured athletes. They quoted one athlete who stated, "When I had that groin injury [the coaches] made me scrimmage anyway. . . . I mean, I had no business being out there" (p. 258). Coaches' attitudes and behaviors toward injury appear to be significant risk factors for OT and reinjury, as exemplified in this quote from a gymnast:

> [Coach] would get mad if I got an injury. He would be so pissed off. He'd be like, "oh no, not again," and then he'd want me in the gym working out and everything. . . . [He] thought that [an injury] was a lack of concentration. So, he was mad at me because, if I was concentrating better, I wouldn't have [gotten injured]. (Krane et al., 1997, p. 59)

As the following excerpt from Krane et al. (1997) illustrates, an athlete may even continue to follow the abusive practices of her coach despite being aware of the negative impact:

> [Coach had an] extremely different concept. This woman, [foreign] born, would place bottle caps on the bottoms of your feet; if you fell on your heels off the balance

beam, then you would have them, the Pepsi bottle caps, go into your heels. [She was] excruciating, die-hard; she was wonderful. You either love her or hate her. I was a person who loved her because she made me so infuriated sometimes and because she was good, and that's why I liked her. (p. 59)

The young gymnast quoted here expressed some major contradictions in the way she viewed her coach—excruciating yet wonderful, infuriating yet likeable. From the description of the coach–athlete situation, it appears that this gymnast was trapped in a dynamic where she would do anything that her coach demanded, no matter how difficult or abusive; it sounds like a situation with very high risk for OT. For more on the dynamics of abusive sport situations, see chapter 6.

Coaches in Conflict

For some athletes, coaches in conflict with one another can add to the stress load already experienced in the training environment: "There was miscommunication between the coaches; coaches were yelling at each other. . . . [I]t was really disorganized and it had a negative impact on me and the other players" (Wrisberg & Johnson, 2002, p. 264). Coaches play important roles in the ways athletes respond to their training environments and make decisions about recovery. The preceding examples suggest that coaches may increase the risk for OT, depending on how they interact with their athletes.

No Downtime

Athletes may feel that they do not have enough, or any, opportunities for downtime away from their sports:

My biggest problem was there was no separation between the role of the father and the coach. So you wouldn't talk about anything else but tennis whether we were eating or if there was a match on television everyone had to watch it, and he'd comment, and whether or not you agreed with him, it didn't matter 'cause, you know, he was always right. You had to do it this way and, you know, he always made us do certain exercises when he wanted and was very strict on getting things done the way he thought, and he didn't leave any room for personal feelings. (Gould et al., 1997, p. 265)

This quote speaks to the situation where the coach is in dual roles (i.e., coach and father). Having a powerful other, like a father, in dual roles can be immensely stressful and confusing for an athlete: "Who am I talking to? Dad or the coach? Who is punishing me?" When the coaching role invades the family life, the stress increases, and the athlete never gets a break but is always vigilant around the father who is also the coach. The increased stress and vigilance in this situation can also lead to stress–recovery imbalances.

Demands of Sport Culture

Another significant stressor for athletes can come from the demands inherent in the sport culture. For example, in figure skating, the constant concern about appearance may cause athletes to feel extra pressure to train harder or experience stress that disrupts their recoveries:

You should do a whole story on weight in figure skating; it is such an appearance sport. You have to go out there with barely anything on. . . . [I]t's not like I'm really skinny or anything, but I'm definitely aware of it. I mean I have dreams about it sometimes. So it's hard having people look at my thigh and say, "Oops, she's an eighth of an inch bigger," or something. It's hard to do. Weight is continually on my mind. I am never, never allowed to be on vacation. Weight is always on my mind. (Gould, Jackson, & Finch, 1993, p. 149)

This type of concern about appearance and weight possibly may lead athletes to push harder in their training and neglect important recovery strategies (especially those related to nutritional intake), thus putting themselves at greater risk for OT and its negative outcomes. As with the athlete whose coach was his father, sports with strong pressures to control weight and appearance may leave the athlete with no downtime.

Stress of Expectations

OT risk might also increase when athletes experience stress or feel pressure from the expectations of others around them. Gould et al. (1993) presented the following excerpt:

Expectations are definitely a concern and they are not a superficial one. Will I measure up to other people's expectations? It is much easier when you don't have any expectations, because if you don't do very well, people just don't notice you; you can always do better next year. But if you do bad with expectations upon you, they condemn you, so that's a stress factor. (p. 147)

Institutional Pressure

In other cases, athletes might perceive excessive pressure at an institutional level:

> The athletic department standards are so high here. The beginning of the season we came home from our first road trip, and we were ashamed to tell people we lost. We felt like we let the whole department down. (Wrisberg and Johnson, 2002, p. 263)

Student-athletes at high schools and universities face not only the stressors of training and competition but also the demands of maintaining grade point averages so they can compete. At universities, scholarship athletes also stand to lose their funding if their academic performance does not meet standards.

More Experienced or Talented Athletes

In yet other situations, athletes might find that the presence of more experienced or talented athletes can create overwhelming stress, as one college swimmer told Wrisberg and Johnson (2002):

> I went to NCAAs [U.S. national collegiate championships] and it was unbelievable the people I saw there. It was huge names in swimming, and I felt so out of place . . . like I didn't belong in the same pool with them. I had a really bad asthma attack, and I think maybe part of that could have been the anxiety. I was completely psyched out. (p. 263)

This quote illustrates that sometimes being in the presence of talented others can produce dramatic reactions (asthma) that further stress the athlete. In team situations, some athletes may find other, more talented team members a constant source of anxiety and stress. That stress, and the depletion of resources it may bring about, can tip the stress–recovery balance into the red.

Stress Outside the Sport Environment

Athletes may also be put at risk for OT by situational stressors outside of their sport and training environments. In some cases, friends can be significant sources of stress: "My roommate is one of those people who seems to need some sort of chaos in her life all the time. I just become a victim of the chaos she needs in her life. I dread going home at night" (Wrisberg & Johnson, 2002, p. 260). Athletes may also experience their relationships with significant others as contributing to the overall stress load: "My relationship with my girlfriend was such a roller coaster. She'd let me in close and then just push me away. It just about drove me crazy" (Wrisberg & Johnson, 2002, p. 262). For some athletes, the family financial circumstances can elevate already high levels of stress: "My family was always under financial burdens, sacrificing everything so I can skate. We didn't have the money, and things are going really bad, and it was like . . . caused a lot of tension, you know" (Scanlan, Stein, & Ravizza, 1991, p. 114). For other athletes, family crises may become overwhelming. Wrisberg and Johnson (2002) provided the following quote from an athlete dealing with a particularly difficult home environment:

> My father had gotten laid off, you know, from his job and everything and that was really tough on my family and all that. . . . [T]hat's when he started his drug use, he started experimenting with the "crack." He and my mother weren't getting along and my sister was just a teenager and had no guidance from my folks at all. It was just awful. (p. 261)

In cases where athletes go home to stressful situations and relationships, their overall stress loads are increased and their capacities for recovery are diminished, contributing significantly to OT risk.

In all of the quotations just reviewed, the expectations of others, or the pressure from comparison with others, added to the stress load from which athletes needed to recover. For some athletes, this added pressure may lead to pushing excessively in training in an effort to compensate for perceived shortcomings and live up to the high expectations placed on them.

Internal Stressors

With respect to intrapersonal variables, it seems that certain attitudes, personalities, and personal experiences may elevate stress levels in athletes, prompt excessive training practices, interfere with recovery, or put them at greater risk for injury.

Supermotivation   Although motivation to train hard may be important in sport, some athletes at risk for OT seem to display supermotivation, wanting so much to succeed that they don't take into account or attend to their own limits. Here is a college tennis player voicing his concern about a teammate's seemingly excessive level of motivation:

> One guy on the team doesn't know when to stop with his training. He's a great guy, but he works too hard, and now he has a stress fracture in his back. Last year he broke a foot. He goes too much. He wants it too bad. (Wrisberg & Johnson, 2002, p. 258)

Concern About Taking Time Off Even though regular recovery should be an integral part of a training program, athletes at risk for OT may express concern about taking time off. This college distance runner expressed the irrational thought that rest is a sign of weakness:

> I feel like I'm weak if I decide to take a day off. It's like, I'm not, you know—I set pretty high standards, you know—if you can't get out there and run, then what are you doing running [NCAA] Division I track? (Wrisberg and Johnson, 2002, p. 258)

Perfectionist Tendencies Many athletes may be described as having perfectionist tendencies, but at the extreme end of those tendencies they may react in maladaptive ways that undermine performance quality. The following comment by an elite figure skater shows how perfectionist thought processes might create frustration for the athlete: "I was a perfectionist. . . . That's probably the hardest thing; I was just a perfectionist all the time. . . . I would never accept myself not doing it perfectly" (Scanlan, Stein, & Ravizza, 1991, p. 115). At the extremes of perfectionism, some athletes will show great dissatisfaction with anything less than winning. Krane et al. (1997) illustrated how a gymnast with extreme perfectionist tendencies could become severely self-deprecating in response to any performance that did not result in a win:

> I would be like, "all of that hard work is down the drain and here you are in 3rd place. You are such an idiot. You are so low; I cannot believe you are here; you are supposed to be up there." . . . I pictured 2nd or 3rd to be, I don't know, what I pictured as a loser. I only knew how to envision a winner because a loser was not even, was not in my world. That picture

may be weird but that's the way it was. It was just tops; everything's always gotta be the top. (p. 63)

Such self-imposed demands for perfection can be a constant source of stress and anxiety, and they can lead athletes to engage in a variety of OT behaviors, such as excessive physical training to compensate for feelings of being a loser.

Self-Destructive "Coping" Behaviors For other athletes, self-destructive behaviors in sport could be connected to poor life choices that increase stress, disrupt recovery, or threaten their health. This college swimmer admitted turning to substance abuse as a coping strategy:

> One year I didn't perform well at all. I felt like I was missing something. I wasn't really part of the team. That's when I started drinking. I don't know, just my personality being kind of obsessive-compulsive. . . . [O]nce I started to drink, I couldn't stop. (Wrisberg & Johnson, 2002, p. 265)

Substance use and abuse is common in sport, and often it is part of the culture (e.g., beer and rugby), but substance use may also be a way of coping with the stress of sport. In time, however, this self-medication becomes yet another stressor in an athlete's life.

Unreasonable Attitudes Toward Training Athletes at risk for OT and injury may also have unreasonable attitudes toward training, sometimes engaging in excessive, unhealthy, or damaging practices. The following quote from a gymnast (Krane et al., 1997) illustrates the extent to which an athlete might go in self-damaging thoughts and behaviors:

> In my mind, practice made perfect. I had believed that pain is gain. . . . And the more it hurt, the better it was; and I don't know why it was like that, and sometimes I think about that, and I think I was becoming psychotic, but I would purposely hurt myself to make myself better. To almost make myself feel like I was existing. (p. 65)

These quotations from athletes suggest that any of numerous stressors and driving factors in athletes' lives may add to their total stress load, prompt poor decisions about training and recovery, and push them over the edge from healthy training to OT. In some cases, it seems that athletes feel pushed by coaches or parents. In other cases, the pressure or maladaptive behaviors may be driven by the athletes' own personalities. Developing a complete picture of athletes' lives seems to be an important step in minimizing the risk for OT.

DIRECTIONS FOR RESEARCH ON OVERTRAINING PHENOMENA

As Armstrong and VanHeest (2002) stated, "the border between optimal performance and a performance impairment due to overtraining is subtle" (p. 187). Such subtlety might prompt exploration at an individual level to gain understanding of the meaning of overtraining experiences for athletes, including why they may risk overtraining or neglect their recovery. Our understanding of risk factors for OT might be enhanced by a holistic approach that encompasses fleshing out the many factors—personal and situational—affecting athletes' lives. Botterill and Wilson (2002) stated that "since the phenomena involved in overtraining and recovery are clearly multifactorial, qualitative descriptive case studies and research can assist us in understanding the complex relationships involved" (p. 143). The literature surveyed here shows that a number of qualitative studies (and several review articles) have initiated investigations generating information about overtraining risk factors. These studies have provided data from both experts' and athletes' perspectives. It seems that the time is right for further in-depth, qualitative research focused on OT phenomena with a view to identifying risk factors.

As the athletes' quotes provided in this chapter have illustrated, and as Gould and Dieffenbach (2002) stated, "it is also evident that researchers must look beyond mere physical training as a cause of overtraining and burnout. . . . Other factors such as psychological stress, inadequate rest, the type of recovery activity, travel, personality, and sociological issues must be examined in multifaceted models" (p. 33). Furthermore, Kenttä and Hassmén (2002) noted that "only the individual athlete knows exactly in which way the training affects her body and mind and how she perceives recovery actions" (p. 67). We suggest, as do others, that it could be useful to conduct more research looking into as many variables as possible in athletes' lives, and providing insight into athletes' experiences from their own perspectives.

Kenttä and Hassmén (2002) noted the difficulties with holistic research approaches: "[I]t may be frustrating to include and consider the individual's whole life situation" (p. 67). Nonetheless, in stating that "performance development and optimal training depend heavily on the ability to integrate and react to as many relevant variables as possible" (p. 67), they supported the rationale for an idiographic approach to examining athletes' experiences with OT and outcomes. Kenttä and Hassmén also stated that "more research is needed in order to help establish to what extent psychosocial stress interacts with training-induced stress in the development of the overtraining syndrome" (p. 73). With such research, one may be able to "increase the understanding of why the same athlete responds differently to a given training stimulus under different conditions, why homogeneous groups of athletes display different responses to a given training stimulus, and why some athletes seem to be more vulnerable to staleness" (Kenttä & Hassmén, p. 74). There is an implicit demand here to understand the individual, a demand that provides the rationale for in-depth studies of individual athletes' experiences with overtraining processes and outcomes. The research we report in this book is about gathering information on and identifying the many relevant variables, available cues, and early warning signs connected to overtraining processes and outcomes.

CONCLUSIONS

Susceptibility to overtraining and potential risk factors have been identified as important areas for research, especially with a focus on individual differences (Budgett, 1998; Flynn, 1998; Raglin, 1993; Uusitalo, 2001), and some researchers have suggested a number of personal and situational risk factors for overtraining, such as coaches and their behaviors, training environments, perfectionism, personality dispositions, and sport cultural imperatives (Gould, Guinan, Greenleaf, Medbery, & Peterson, 1999; Gould et al., 1997; Krane et al., 1997; Uusitalo, 2001). There is no published research, however, conducted systematically to uncover, understand, and test such potential risk factors. Currently, many top coaches and athletes appear to have access to sport science knowledge that may enable them to minimize or avoid OT. Nonetheless, many athletes are still pushing too hard and are at risk for serious OT outcomes. Idiographic research directed toward revealing what is experienced at the level of the individual athlete could be helpful in presenting a more complete understanding of the OT process, including the risk factors, actual causes, and potential consequences.

It is important to identify predisposing personal variables, as well as situational variables, that are present before a particular training cycle commences. It might then be possible for athletes, coaches, physicians, and sport psychologists to be more sensitive to risk factors and to act to minimize negative outcomes. Raglin and Morgan (1989, 1994) found that, among college swimmers, 91% who became stale in their first season also became stale in one or more subsequent seasons, whereas only 30% of the swimmers who did not become stale during their first season developed staleness in another season. Raglin (1993) concluded that this finding "suggests that athletes display consistent differences in their propensity toward becoming stale by the age of 18 and indicates that some individuals are at greater risk of suffering the disorder than others" (p. 843). O'Toole (1998) stated that individual variability in responses to a given training load is such that a particular exercise program may be optimal for one athlete yet lead to overtraining syndrome in another. It could be important to ask what the sources of this variability are. Identifying which athletes are at higher risk for OT and its negative outcomes could be one of the first steps toward prevention. Perhaps with more research on risk factors, it might be possible to predict what situations are most likely to lead to overtraining and what types of athlete are most likely to overtrain.

WHAT THE EXPERTS HAVE TO SAY

In this section of the book, we consider the experts' perspectives. As with the views of athletes, we used interviews to explore experts' experiences of working with athletes going through overtraining processes, sometimes to the point of experiencing overtraining syndrome. Initially, we interviewed 14 experts from major sports organizations in Australia. In chapter 4, we summarize the rich and detailed information about potential risk factors for overtraining that we gleaned from these interviews.

The 14 experts we interviewed included four national coaches, four sport physicians, five sport psychologists, and one sport physiologist. We chose them because they have substantial experience in working with elite athletes who have overtrained. Initially, we made contact with them through their organizations. They signed standard consent forms before their interviews. For ethical reasons, their identities remain confidential (except for Dr. David Martin and Dr. Trisha Leahy), and we made every effort to ensure that no material was reported that could identify any athletes or experts. We stopped collecting information when the stories told in the interviews stared to become repetitive with earlier interviews and did not generate any new ideas. The interviews were open-ended to allow the experts to tell us stories, in their own ways and in their own words, about their experiences with athletes who overtrained. We had in mind our general questions about risk factors for overtraining, and we encouraged the experts to present stories related to risk factors.

Making sense out of the experts' stories involved a number of stages during which the content of the interviews was grouped into comments that related to the same point. For example, we noticed that a number of statements made by experts referred to athletes who seemed to have "supermotivation." This grouping appeared to us to go together with other groups of expert statements that related to extreme perfectionism, obsessive-compulsive characteristics, and strong athletic identities. We propose that this larger grouping is about personality factors. We then grouped personality factors with a number of other categories related to the athlete, as opposed to aspects of the environment. These personal aspects are: behaviors, attitudes, and experiences of athletes. We call these broad groupings *general dimensions*. In chapter 4, we summarize the groups that made up each of the three general dimensions that we identified.

Despite the length of the interviews (1 to 1.5 hours), all the experts' stories of overtraining contained material that we would have liked to explore further, but we decided that it was impractical to go back to all of them. Instead, we identified two experts (Dr. David Martin and Dr. Trisha Leahy) whose comments particularly intrigued us, and we interviewed them in depth again. We studied these new interviews repeatedly and read them in light of the original interviews with these two experts. Then, in chapters 5 and 6, we wrote detailed accounts of these two experts' views, presented primarily in their own words, with occasional comments from us.

The chapters in part II of the book reflect quite different ways of telling stories about experts' views of overtraining. Chapter 4 is a summary, in narrative style, of the analysis of a large body of interview material provided by 14 experts. It establishes a starting point for a framework of risk factors. Chapters 5 and 6 provide readers with the opportunity to enter two experts' professional landscapes and to appreciate quite different, yet equally valid, views of athlete overtraining, given by people who clearly feel strongly about the forces that contribute to the phenomenon. Let's now move into the world of experts in sport and exercise and discover what they think leads athletes to overtrain.

CHAPTER 4

COACHES' AND SPORT SCIENTISTS' VIEWS ON RISK FACTORS

In their interviews, 14 experts gave us a large amount of rich and valuable information about the characteristics and situations they perceived to be related to risks for overtraining (OT). Many of the experts confirmed what we had found in the literature covered in chapter 3, but in several instances the experts went far beyond what has been published and shed more light on possible risk factors. Their stories made the extant literature come alive and added to our understanding of risk factors. To make sense of this vast amount of information, we transcribed the interviews, grouped the material into common themes, then brought together themes that reflected a common issue. To reduce the chance that the emerging framework might represent the expectations of one person, at least two of us independently did the grouping so we could compare themes and reach agreement through discussion of cases where we had arrived at different groupings. Luckily, we found a large degree of congruence.

From the rich information provided by these experts—coming from their own varied perspectives as coaches and medical or psychological support staff—a picture emerged of many risk factors for OT. After several rounds of analysis, we arrived at a way of grouping these factors that classified them into three broad categories or general dimensions:

- Characteristics, behaviors, attitudes, and experiences of susceptible athletes
- People, factors, and situations that pressure athletes to increase training

- People, factors, and situations that affect athletes' needs for recovery

Within each of these general dimensions we found a number of themes, and within those themes we identified various issues. Almost all the issues reflected comments made by several experts; in some cases, most of the experts commented on a particular issue. In Sean's project, he initially allocated around 100 pages to the presentation of this material, supporting each issue with word-for-word quotes from experts. Yet those quotes were still only a sample reflecting the experts' voices. Considering the aims of this book, we can paint only a broad-brush picture of this material; we will summarize the themes and issues within each dimension in a much shorter version. Readers who wish to examine the experts' statements in more detail should contact us through Human Kinetics. We will be happy to send more information.

CHARACTERISTICS, BEHAVIORS, ATTITUDES, AND EXPERIENCES OF SUSCEPTIBLE ATHLETES

It is hardly surprising that experts perceived that the potential to overtrain is related to a variety of phenomena making some athletes more susceptible than others to overtraining. Within this general dimension, we identified five major categories of athlete-related risk factors:

- Character or personality
- Physical susceptibilities
- Personal experiences
- Personal beliefs, attitudes, and expectations
- Specific behaviors

Character or Personality

Experts saw no specific profiles of athletes susceptible to overtraining but did identify five traitlike characteristics that seemingly predispose athletes to go overboard:

- Obsessive-compulsive characteristics
- Extreme perfectionist characteristics
- Supermotivation, or extremely high internal drive
- Extremely strong athletic identity
- Other psychopathologies

A feature common to all these characteristics is that their manifestation in susceptible athletes is pronounced. These factors are not necessarily separate from each other; they often overlap. For example, some athletes are obsessive about getting every aspect of their preparations exactly right (obsessive-compulsive), but no matter how well certain athletes perform, it is not good enough because it is not perfect (perfectionism). Some athletes seem to be driven to go beyond reasonable limits to achieve more; others see themselves only as athletes, and their lives are totally focused on their sport.

Physical Susceptibilities

Experts pointed out that there are risk factors related to the physical characteristics of athletes. These factors appear to involve age, developmental level, and immune system functioning. The experts noted that some athletes risk overtraining because they undertake activities or levels of activity that are unsuitable for their developmental stages. Sometimes young athletes, whose skeletal and muscular structure is still developing, try to emulate the training and performance regimens of the champions they see in their sports, who are usually mature adults in their physical prime. Coaches may also lead young athletes into damaging practices by training them as if they were adults. At the other end of the scale, older athletes, whose bodies are beginning to lose their capacity for rapid recovery, continue to adopt training and

performance schedules that do not allow adequate rest for return to full functioning. The experts also expressed the view that some athletes develop compromised immune systems, due to their heavy training and performance regimens. This compromise makes them susceptible to illnesses such as minor infections and viruses. Often they continue to train with these conditions, exacerbating the problem.

Personal Experiences

The experts we interviewed proposed that some experiences athletes have during their careers could affect their risk of overtraining. We classified these experiences into three categories: problematic health issues, recent successes or performance peaks, and lack of guidance or adequate support. In terms of health issues, experts referred to increased risk of recurrences of overtraining in athletes who have previously overtrained or experienced illnesses or injuries. Experts saw athletes especially at risk when they had previously received positive reinforcement for maladaptive responses (high praise for thrashing themselves in training) or been abused, whether in or out of sport. In addition, some athletes desperately want to please the coach by working harder and without complaint despite illness or injury, and they may hide their physical and emotional difficulties until it is too late, and the damage is done.

Oddly enough, the experts indicated that even the experience of success can be a risk factor for overtraining. They noted that sometimes performance peaks motivate athletes to try even harder, in the belief that they will achieve still-higher peaks. They also pointed out that early success can lead athletes to develop high expectations, perhaps prompting them to train excessively in order to meet those expectations. Although athletes in peak form might be encouraged to train even harder, the experts observed that they are likely to be precariously close to overtraining, and pushing them further may send them down the slope on the other side of that peak.

Where athletes lack the experience to make sound judgments about their training and performance schedules, they need informed advice and guidance. Unfortunately, according to the experts, coaches and parents—the people closest to young athletes in particular, and often the most relied on by those athletes—sometimes promote overtraining behaviors rather than advising against them.

Personal Beliefs, Attitudes, and Expectations

Another major category of person-centered risk factors for overtraining involves athletes' beliefs, attitudes, and expectations. The experts proposed that what athletes think and understand about training and recovery influences the decisions they make. We classified these cognitive aspects of overtraining into three groups:

- Beliefs and attitudes that extra training is the route to success
- Lack of knowledge or awareness regarding the importance of recovery
- Unrealistic expectations

Experts observed that some athletes believe there is no excuse to miss training and that missed training sessions must be made up, regardless of the circumstances. As a result, when they miss a training session, they believe that they must do that training the next available day, as well as the training already scheduled for that day, thus pushing themselves even harder. Sean and Mark have seen similar beliefs in ballet dancers. One day of missed dance practice means you are two days behind. That belief is an accepted rule in ballet. According to the experts, athletes often believe that more training is always better than less and behave accordingly, even when coaches and others advise them to adopt more moderate training.

Although sport science knowledge is growing in the areas of training and recovery, many athletes and coaches still have not been educated about the risk of overtraining or about ways to reduce the likelihood of its occurrence. Athletes who are not aware of the risks and implications of overtraining are more likely to perceive recovery time as a lost opportunity to train. In the experts' views, such athletes might see rest days as counterproductive.

Athletes often have lofty aspirations based on unrealistic expectations of their current capabilities, and the experts say that aiming too high can lead athletes to train too hard or deny themselves adequate time to recover. As mentioned earlier, one-off or unexpected success can lead athletes to push themselves even harder, because they begin to believe that they can repeat that outstanding performance if they just work hard enough.

Unrealistic expectations, believing that extra training brings success, and lack of knowledge regarding the importance of recovery are some of the reasons athletes may overtrain.

Specific Behaviors

Not surprisingly, the 14 experts also identified certain behaviors associated with increased risk of overtraining. Inevitably, thoughts and behaviors are closely linked, so it is difficult to separate them. We propose two categories of athlete behavior that experts considered to be related to the potential for overtraining:

- Risky behaviors related to doing too much
- Risky behaviors surrounding eating and nutrition

Most experts we interviewed have observed that athletes at risk of overtraining tend to do extra training or push themselves to do more than others in any way they can.

According to the experts, some behaviors associated with overtraining risk are easily observable, whereas others are difficult to pick up. A simple example is that athletes who always stay after regular training, continuing to work on their skills or conditioning, are visible to coaches and other athletes, as well as to those sport science staff who work at training venues. On the other hand, athletes who leave at the end of training sessions but do extra work on their own (and don't report this activity to coaches or conditioning experts) are difficult to identify and, thus, to advise.

Experts particularly pointed to the area of eating and nutrition as being fraught with risks for overtraining. Athletes doing everything they can to be their best may engage in risky eating behaviors, especially in sports where weight, appearance, or physique are critical. Athletes may indulge in questionable nutritional strategies, drastic weight loss practices, and disordered eating, which some experts believe can increase the risk of overtraining. Athletes may also start training harder because they see a need to reduce weight, improve their physical appearance, or gain a performance advantage (e.g., moving to a lower weight category).

In this general dimension (characteristics, behaviors, attitudes, and experiences of susceptible athletes), we have described a wide range of personal risk factors for overtraining. It is important for us to reiterate that the experts stated that each of these factors has the *potential* to lead to overtraining but made no claim that any particular pattern or profile of personal risk factors is typical of most athletes who overtrain. In general, we understand that the experts contended that athletes who demonstrate a larger number of these factors are at greater risk, but that any extreme personality characteristic, attitude, or behavior can lead athletes to overtrain.

PEOPLE, FACTORS, AND SITUATIONS THAT PRESSURE ATHLETES TO INCREASE TRAINING

In the early stages of our delving into the rich and informative material provided by the experts we interviewed, we conceptualized the risk factors for overtraining under two general headings. One general dimension included the personal characteristics, experiences, and behaviors of susceptible athletes, as just discussed, and the other involved situational and environmental risk factors. As we

reviewed and reexamined the experts' statements, we came to perceive that the situational risk factors fell into two broad categories: those aspects that experts argued often push athletes to train too hard, and those that affect the extent of recovery activity that athletes require. Issues related to training and to recovery are frequently intertwined. For example, during the lead-up to major competitions, such as the Olympics or World Championships, athletes may feel motivated to do more than the programmed amount of training. At the same time, they may experience stress due to anticipation of such key events in their careers, for which they need extra recovery from training. We acknowledge the close relationship between training and recovery, but for this part of our argument we have assigned situational risk factors proposed by the experts to either the group that pushes athletes to train more or the group that encourages them not to meet their needs for recovery.

We will look next, then, at our second general dimension—the many people, factors, and situations that the experts discussed as pressuring athletes to increase training. They fall into four major categories:

- Behaviors and attitudes of coaches
- Behaviors and attitudes of family and others
- Specific sport factors
- Sociocultural factors

As the names of these categories suggest, the factors that the experts proposed range from those specific to individuals in the athlete's world, through small group influences, to pressures imposed by the subculture pervading a particular sport or even by the culture of a whole society. We examine each of these categories in turn.

Behaviors and Attitudes of Coaches

The experts we interviewed reported many ways in which coaches' behaviors and attitudes might lead athletes to train to an excessive degree. These risk factors include coaching experience, external pressures on coaches, and coaching style or focus.

According to the experts, some coaches lack experience in overtraining problems and therefore may unknowingly load athletes with too much training. In other cases, coaches might not appreciate that senior (older) athletes, and young athletes just coming through, cannot train at the same level as athletes in their physical prime. But not all heavy training loads come from coaches' lack of experience. Some experts observed that coaches who have tasted success, and perhaps as a result gained status and a high profile, might push athletes too hard in order to sustain that success and its perks. Success can positively reinforce the more-is-better approach, if a coach pushes athletes hard in training and even a few of those athletes are then successful.

Experts also observed that coaches of athletes or teams at the elite level might feel pressure from sport organizations, fans, the media, parents, or sponsors to deliver successful performance. These pressures can lead coaches to push athletes harder. Once again, this push could be based on a more-is-better philosophy, because it is easy to sell oneself on the idea that high-volume training must lead to higher levels of performance.

Coaching style is another aspect of coach behavior that experts argued could lead athletes to push too hard. Coaches have a powerful influence on athletes. If a coach espouses a more-is-better philosophy, it often won't even be necessary for the coach to directly control athletes' training schedules. Athletes hearing the "train hard" message will push themselves to the limit because they know that is what the coach expects. Experts also observed that some coaches adopt a short-term-goal approach, perhaps being as shortsighted as to react to single performances. This approach can lead coaches to drive athletes to train too hard. Some experts also pointed out that autocratic coaches—those who expect athletes to follow their orders and who will brook no discussion—are more likely to overtrain athletes than are democratic coaches, who discuss with their athletes the impact of their decisions about athletes' training schedules.

Behaviors and Attitudes of Family and Others

Although the influence of coaches on athletes is often very strong at the elite level in sport, the athlete and coach do not live in a vacuum. Other people can encourage athletes to train at excessive levels, either intentionally or indirectly. We identified two categories of behaviors and attitudes of family and other people linked to athletes that experts claimed have the potential to affect how much athletes train: parents' and others' emotional reinforcements, and parents and others (e.g., personal managers) who seek personal or financial gains.

Experts reported that some athletes display extremely strong work ethics derived presumably from a family climate that places much importance on working hard. In other cases, experts feel that athletes have been driven by their parents, who might be ambitious for the athletes to achieve in ways they themselves have not. Within the context of reinforcement by family and others, experts again raised the issue of the more-is-better philosophy. They argued that others often positively reinforce athletes. Family, friends, teammates, fans, the media, and, according to some experts, even some psychologists can praise hard-training athletes, making them think that they are doing something good. Some experts argued that athletes who sought parental love or approval might train harder and harder in an effort to elicit positive reactions from their parents. Where parents are overinvolved in their child's sport, experts proposed that their example might lead an athlete to train harder. One expert also commented that athletes' parents may encourage athletes to train even though they are ill or injured, rather than supporting proper recovery, because the parents are worried about the training time being lost. This factor overlaps other categories and could also be placed in the general recovery dimension.

In many sports, the personal and financial rewards for success are large, and sporting environments are often highly competitive. Sometimes the winner gets a lucrative contract or sponsorships whereas the person who comes in second is quickly forgotten. In these circumstances, parents, partners, managers, agents, and others can place extra pressure on athletes to train even harder, in an attempt to ensure that the athletes gain financial and personal benefits, some of which they anticipate will come their way. Although experts acknowledged that parents and others might feel genuinely proud of their high-performing child, they also referred to occasions where parents seemed to be compensating for their own lack of success, living vicariously through their children's success.

Specific Sport Factors

The third major category of risk factors perceived by experts in coaching and sport science involves issues directly or specifically related to sports that push athletes to increase their training to excessive levels. We classified the sport-specific factors into five groups:

- Pressure to gain financial reward or support
- Timing or scheduling factors
- Physical factors, such as age or weight
- Sport- or training-specific environment or culture
- Transitional factors

Money pressures affect most if not all sports. Experts noted that athletes can be motivated to train at high levels by the pressure to earn high professional salaries or win large monetary prizes. They also noted that athletes might feel great pressure to live up to massive salaries or prize money after they have been acquired. In addition, experts pointed to the pressure inherent in salary systems that include bonuses for each game played and, often, still bigger bonuses for winning. Similarly, it is often difficult for athletes to sit out high-paying tournaments or other events even if they feel fatigued. In other sports, the financial rewards are not so large, and in many such cases the athletes are dependent on government grants or other benevolent funding to allow them to train at a level sufficient to be competitive. Experts suggested that when such athletes feel that continued funding depends on results, they are at risk of training too hard to try to deliver them.

Experts pointed out that many sports involve certain times during a season when pressure to train hard is particularly high for athletes, such as competing for a place on a team for an Olympics or a World Championship. At such times, athletes might push on through injuries and fatigue, thus, ironically, putting themselves at risk of overtraining right before, or even during, the event for which they have been working so hard. Experts claimed further that, even after being selected, athletes remain at risk of training too hard in the lead-up to major competitions. Once again, this problem stems from the belief that training hard, or harder, is the best preparation for major events, so the looming event provides the trigger. Experts suggested that the push to train harder at these key times could be associated with increases in illness or could produce a downward spiral of increasing training and worsening performance.

Some sports also bring pressure to train hard in order to attain a desirable body weight or shape, whereas other sports demand high performance that is dependent on heavy training at young ages. Experts reported that these pressures due to physical factors could drive athletes to train especially hard. One common situation that experts discussed was the impact of skinfold monitoring on training intensity of athletes. The view usually is that skinfolds need to be as low as possible. Experts pointed to sports like gymnastics, where athletes are expected to perform at the highest level when very young, possibly pushing them to train excessively at a vulnerable stage of physical and psychological development. In professional Australian Rules football, skinfold targets are sometimes written into contracts, even though the relationship between skinfolds and performance in Aussie Rules is tenuous at best. Every sport psychologist we know who works in Aussie Rules football estimates that at least 10% of the professional players struggle with pathogenic weight control.

Experts also raised the issue of the environment or culture of particular sports (perceived as traditions, customs, or just the way things are done) that athletes are expected to follow, regardless of their situations. One expert argued strongly that the whole elite sport system is largely responsible for overtraining and that shifting the focus to individual factors amounts to "depoliticizing" the issue (see chapter 6). Another expert noted that athletes could be put at great risk of overtraining by those who follow traditional approaches to training (because that is the way it has always been done) without examining the potential of those practices to cause harm. Experts also described sports in which selection at the elite level is based on a Darwinian principle of survival of the fittest, especially where there is a large pool of athletes training at high levels, allowing coaches to simply keep pushing them harder because there will always be those who can handle the training without getting injured or overtrained. Many athletes are at high risk of overtraining in such systems, but it is usually the successful athletes who gain media attention, and the impression is that the training methods are working well.

Another way in which experts claimed that people close to athletes, or even the athletes themselves, may encourage overtraining is through comparison of the training and performance of different athletes. This problem can also be systemic, when, for example, it is customary to put lists of performances at training and competitions on a notice board. A different version of the comparison process occurs when there has been an outstanding athlete in the past, whose vital statistics are presented to all current athletes as the gold standard. One expert stressed that all athletes have different needs, so trying to produce another Lance Armstrong or Ian Thorpe is likely to be fraught with risk. Experts also suggested that being a competitive athlete in itself seems to involve implicit demands to overtrain. They argued that athletes often do not know their limits, so they might often need to overtrain, or become ill or injured, before they realize that they have pushed too hard. Experts proposed, however, that when athletes have previously become injured or overtrained but don't learn to recognize their limits, they might be at greater risk of overtraining again. See chapter 8 for a story of failing to learn limits and, as a result, engaging in repeated overtraining.

As athletes develop, they move into different phases of their careers—from promising or high-ranking juniors, to rookie senior players, to mature players, then into the twilight of their careers and, finally, out of competitive sport altogether. Such transitions are accompanied by factors that can increase the risk of overtraining. Movement into different phases (e.g., from part-time to full-time training, as often happens when players join professional clubs or receive stipends from sport institutes or national training centers) often involves increases in training frequency and intensity. Another kind of transition occurs when an athlete who has been injured for a substantial period of time begins training again. Such athletes may assume that they can start to train at the same level they were on when they became injured, and this kind of thinking and action carries a high risk of overtraining. Even if they know that they should start at lower levels of training and build up gradually, the frustration of sitting on the sidelines, and the fear of not regaining their previous level of fitness and skill, can be powerful drivers. As noted earlier, athletes nearing the ends of their careers might be pushed by thoughts of retirement to try to train at the same level as younger teammates, who are at their physical peaks, or they might try to maintain the training levels they operated at when they were younger. We have all observed many instances of such athletes pushing themselves too hard as they try to hang on to the activities that have given them their identities for many years. Ironically, if they don't sustain career-ending injuries first, these athletes are increasing the risk that overtraining will end their careers.

Sociocultural Factors

The final major category of risk factors in this general dimension (increasing training) involves sociocultural influences. This category refers to attitudes, norms, and imperatives imposed by

Sports like the triathlon, for example, are built on the idea of toughness, which can lead athletes to train beyond their capabilities.

the sociocultural environment that can influence athletes to increase their training to excessive levels. We identified two groups of factors in this category:

- Reinforcement for attitudes and beliefs that support overtraining
- Reward for pushing very hard in training

Athletes can be influenced by societal norms and pressures in sport or by accepting situations that favor overtraining. One expert in particular emphasized that sport is often organized hierarchically and that coaches have enormous power, so the system accepts high levels of abuse of athletes. This expert observed that often parents do not intervene, even when coaches are openly abusive to their children, giving a message to the children that the process is "normal" and that people are allowed to abuse them. This situation can lead to athletes developing dysfunctional coping strategies such as overtraining or pushing so hard that they get injured (see chapter 6). A number of experts also pointed out that the "more-is-better" mantra is common in much of society (not just sport), thus reinforcing athletes' experiences of it in their sport. An expert also noted that further pressure often comes from the media, which treats the injured athlete who plays on as a hero. Often, the media also reflect the notion that a given player is indispensable to the team, placing additional pressure on athletes to play when they should be recovering from injury or taking a break because they are exhausted.

Rewards for pushing hard can also come from national pride in sport achievement, which might be associated with financial reward. Experts noted that, in some countries, success in sport raises not only athletes but also their families out of poverty. In some cases athletes become wealthy and can support their families, whereas in other situations the state directly looks after the families of elite athletes. Experts also noted that cultures often promote the view that sport success has a big influence on national morale, so athletes feel that they must do well to lift the morale of the people, driving them to train at harmful levels.

In this general dimension of people, factors, and situations that affect athletes' drive to train excessively, we have seen how factors at many levels can influence athletes' behavior, leading them to overtrain. These variables range from specific individuals who are close to athletes, such as coaches and parents, to the most general attitudes and norms of the society, promulgated by impersonal mechanisms such as the mass media, including television, radio, the Internet, newspapers, and magazines. Many of the same mechanisms also operate to limit the amount of recovery that athletes experience during demanding schedules of training and competition. In addition, however, there are stressors that change the amount of recovery athletes need, often without recognition or modification of recovery schedules, and this issue is covered in the next section.

PEOPLE, FACTORS, AND SITUATIONS THAT AFFECT ATHLETES' NEEDS FOR RECOVERY

This general dimension refers to a range of variables that influence the overall stress load with which athletes must cope, usually raising the total load. These risk factors do not push athletes to train harder but do create extra stress, increasing the need for recovery. From the rich reports of the experts we interviewed, we identified four major categories of risk in this dimension:

- Behaviors and attitudes of coaches
- Behaviors and attitudes of family and others
- Specific sport factors
- Other life factors

Behaviors and Attitudes of Coaches

The experts described a number of behaviors and attitudes of coaches that might create a need for extra recovery or deny adequate recovery time. We include here only factors that do not pressure athletes to do extra training, yet do increase the demand for recovery, and we have grouped them into two categories: those related to lack of knowledge, understanding, or awareness, and those related to health issues.

Just as coaches usually wield the greatest influence over training, they also have a powerful influence on recovery. Experts pointed out that when coaches have shortcomings in knowledge, understanding, or awareness of risky situations, the probability increases that athletes will overtrain due to inadequate recovery. Experts suggested that some coaches do not appreciate that athletes' total stress includes factors outside their sport. According to the experts, personal factors add to the overall psychological stress experienced by athletes, thus influencing their recovery needs. Experts also cited some coaches' lack of knowledge of sport science. For example, some coaches still believe that athletes should train hard right up to an event; they are unaware of the need for tapering, or easing the training in the build-up to major events, which is now well-established in the sport science literature. Experts also noted that it is important to appreciate the physical and psychological differences between athletes. Experts claimed that some coaches adopt a one-size-fits-all approach, rather than tailoring training and recovery to fit the unique needs of each athlete.

Experts argued that coaches are also often unaware of the ways in which health issues affect training and recovery needs, so they simply do not address those concerns. A prime example of this problem is that coaches are frequently unaware of the impact of reinforcing sociocultural norms that are prevalent in many sports. Those norms (e.g., "don't complain," "play on when injured") help silence athletes in the face of injury, illness, and fatigue. Further, the experts claimed that, whether explicitly or implicitly, coaches often encourage athletes to return early from injury or illness.

Behaviors and Attitudes of Family and Others

Experts also proposed that family and other people who interact closely with athletes, such as athletic support staff, may exert influences that increase athletes' recovery needs or interfere with athletes' recovery processes. We divided the behavior and attitudes of family and others into attitudes toward life balance and other stressors and factors related to health issues. Because athletes appear to function effectively in the sport environment, experts argued that many people who provide services for them may assume that athletes also function well in most other spheres of life. If family and service staff do not understand that they need to provide support for the whole person, there is a risk that they will not appreciate the other stressors being experienced by athletes and, as a consequence, underestimate athletes' needs for recovery, which is likely to increase the risk of athletes overtraining. In relation to health issues, experts suggested that family and others close to athletes might not recognize the seriousness of health problems that athletes are experiencing. Perhaps because of a real desire to provide support, these people might encourage athletes to believe that they can continue to train or perform when the athletes have "sniffles" or "niggles," not appreciating that such minor conditions can develop into more serious illnesses or injuries. Experts stated that even medical staff can approve athletes' premature return from injury or illness in an ill-conceived effort to support them.

Specific Sport Factors

We identified four groups of factors suggested by experts that are directly related to sports:

- Training program design
- Pressures from financial strain or lack of resources
- Transitional issues
- Other sport-related factors

Experts considered the design of training programs to be one of the most important factors affecting training-and-recovery balance. Several experts indicated that highly repetitive, high-volume training schedules can affect recovery because they constantly place stress on the same parts of the body or the same energy systems without giving adequate time for recovery. Repetitive schedules are also likely to increase psychological fatigue from which athletes need time to recover. Experts also noted that an overemphasis on one area of training can present problems for recovery. They noted that athletes tend to favor their personally preferred types of training when they do sessions. For example, they might prefer strength over conditioning work. The experts advised that athletes must balance different activities to minimize the risk of overtraining. If athletes keep working on their quads, the demands on those muscles increase and more recovery is needed for them. In addition, it is possible for athletes to lead coaches to underestimate recovery needs if they use a number of training advisors with no overall coordination of what they do and no communication about it with the coach. A similar problem might arise in multidiscipline sports, where each athlete might have different support for each discipline. If nobody has the whole picture, there is greater risk of underestimating the recovery needs of such athletes. Finally, in this category, an expert noted that training programs based on maintaining peaks in performance are likely to leave insufficient time for recovery, increasing the risk of overtraining.

Experts suggested a number of ways in which pressures from financial strain or lack of resources might reduce recovery time or increase recovery needs. If athletes do not get sufficient support for the equipment, training, and travel associated with competing, their recovery needs are likely to be greater when the time available for recovery is shorter. To make up the shortfall, athletes might take on work or sponsorships that eat into recovery time and, due to the stress they bring, increase recovery needs. If athletes decide not to work to meet financial shortfalls, then they are likely to experience extra stress because of increasing financial pressures. An expert also noted that athletes with financial difficulties might not be able to afford to pay for important support services or resources. Thus it seems that, in terms of finances, many athletes cannot win—either they overstretch themselves or they miss out on important support.

During transitions, athletes are likely to experience additional stress. Experts reported that such stress requires extra recovery time that often is not allotted. In the case of stress associated with entering higher echelons of sport, it is probable that athletes have less time for recovery than they are used to, even as they experience more stress. In addition, stress can be increased by changing one's physical environment, switching time zones, moving to a different altitude, or performing in, say, a hot rather than a temperate or cold environment. Those who plan an athlete's training often do not take into account such stressors by assigning the needed additional recovery opportunities. One expert noted that recovery needs might also be underestimated during tapering for major competition. Allowance might be made for the taper itself, but planners may forget that stress is likely to increase because the major competition is getting close.

The experts also raised sport-related issues connected to athletes' recovery needs that did not fit into the previous categories. For example, risk is increased when a sport has a lengthy season or high frequencies of competition, which is a common situation in professional sports. Some athletes in sports that are played seasonally in the northern and southern hemispheres—such as basketball, cricket, and rugby—have no off-season at all, going directly from a league season in one part of the world to competitions on the other side of the planet. Experts also noted that in sports where high training volumes are the norm, not only are the risks of overtraining increased, due to heavy training and insufficient recovery, but also the identification of overtraining can be hindered because everybody is training at such high volumes.

Other Life Factors

The final major category of variables affecting athletes' recovery needs includes factors that do not fit into the other categories. Broadly, these factors originate in areas of athletes' general activity and commitment outside of sport, and they influence athletes because they demand time or emotional resources.

One kind of commitment, which experts noted requires a substantial investment of time and can

also be stressful, is involvement in school or university study. As well as spreading themselves thin, athletes studying for degrees face the stresses of school assessments and assignment deadlines. As experts pointed out, recovery from training for athletes in these situations is not the kind of relaxing activity indulged in by many professional athletes; it is often a case of rushing to the library or the next lecture. Some experts also noted that, especially as they mature, athletes also have to manage the kinds of life stressors that most adults address on a daily basis. They are faced with concerns such as marriage, families of their own, setting up and moving house, handling mortgages and other major financial commitments, and family illnesses. An athlete who trains at 5:30 in the morning after a night disturbed by an unsettled baby is likely to have different recovery needs than the residents at a sport institute or national training center, who have accommodation, food, and training facilities provided for them. For higher-profile athletes in particular, experts noted that time for rest and recovery can also be disrupted by media appearances and publicity commitments, such as opening shopping centers and sporting goods stores.

CONCLUSIONS

In this chapter, we have described our analysis of the rich information that 14 experts provided about overtraining, and a number of general conclusions stand out from this material. First, although it was not unexpected, we must note the huge range of risk factors for overtraining generated through the interview process. All the points made by the experts appear to be real issues that many athletes struggle to manage. Second, the variety of variables that experts raised speaks to the need for some kind of model or framework to help us to conceptualize the risks, let alone address them. Third, we note that there was a clear distinction between factors increasing the likelihood of training excessively and factors affecting athletes' needs for recovery, as well as the time available for recovery activities.

We consider that the exploration of experts' perspectives on what leads athletes to overtrain has proved fruitful. We have identified a large number of risk factors and placed them in a framework. Examining the parameters of various risk factors and the utility of the framework are, we suggest, important directions for future research. At the same time, we recognize that the approach used in the present exploration only skimmed the surface. Thus, we decided to dig deeper into the perspectives of two experts, and we report their stories in detail in the next two chapters. Furthermore, the picture that emerges from the interviews with the full group of experts, though rich in terms of the number of risk factors generated, is fragmented, because the factors come from experiences with a wide variety of athletes, each facing different circumstances. Thus, we also examine overtrain-ing stories of individual athletes to try to build up illustrative pictures of overtraining. Those stories are found in part III of the book.

For now, we say thank you for the valuable contributions of the 14 experts who gave us their insights into risks for overtraining. We move on to explore in much greater detail the views of over-training in elite athletes that were presented by two of the experts whose opinions and insights were especially well developed and cogently argued.

Burnt Cookies

Conversations With an Exercise Physiologist

In the previous chapter we presented stories from experts in the field (psychologists, exercise physiologists, and coaches) about their experiences with overtraining (OT). Those stories are in a *realist tale* format (Sparkes, 2002) and give a broad picture of overtraining. For this chapter, we present, in depth, the experiences of an eminent practitioner, an exercise physiologist who has worked with athletes at the highest levels.

INTRODUCING THE EXERCISE PHYSIOLOGIST

David Martin, PhD, has been employed for over a dozen years at one of the world's premier sport training facilities, the Australian Institute of Sport (AIS) in Canberra. Dave has a wealth of experience with athletes who have overdone it. He is also one of the best lay psychologists we know. He has a keen appreciation of human frailty and of the motivations that drive athletes into training that ends up damaging them. Dave's story, of course, starts with Dave. Sean and Mark both interviewed Dave. Sean's conversation with Dave revolves around risk factors for overtraining at the global level (overtraining syndrome) and the local level (e.g., overtraining of a joint or muscle group). Mark's discussion with Dave explores the relationships that exercise physiologists have with athletes and coaches, along with Dave's views of the processes involved in delivering exercise physiology service. This chapter and the following one contain primarily the dialogues with these experts (edited for clarity).

STUDYING ONESELF

As psychology students begin their academic and professional education, it is a nearly universal phenomenon that, in some ways, they are entering the field to study themselves and make sense of their worlds. We suspect that this motivation to understand oneself is also present in many other fields of study outside of psychology. Dave was an elite athlete, and his experiences with the ups and downs of his physiology and his performances fascinated him and were part of his motivation to become a physiologist. Like many psychology students, Dave wanted to understand his body, his training, and the mysteries surrounding athletic performance. So let's start with Dave, in his own voice, giving us some background on how he ended up as an exercise physiologist at the AIS.

David Martin (DM): I grew up with a dad who loves skiing. He would only take a job where there were ski areas, in Idaho, Wyoming, and Oregon. If we moved, it was to another place where there was a good ski area. So we started skiing quite young; my brother and sister and I were all competitive downhill skiers, and when I was a senior in high school I won a medal at the Oregon state championships. Eventually, I got a ski scholarship to go to the College of Idaho, so I went over to Caldwell, Idaho, and tried to improve my downhill skiing. I had always cross-country skied, ever since I was little, and I jumped into some collegiate cross-country races, thinking I was just doing it for training. I was doing very well, so my last 2 years I was a combination skier or skimeister. In my junior year, I became the NCSA [National Collegiate Ski Association] skimeister champion. For my senior year, I only competed as a cross-country skier. Then I started really getting into cross-country skiing; I placed top 20 at a U.S. National Championships and started thinking maybe I should be an Olympian cross-country skier. But I had just finished my undergraduate studies, and I really didn't know what I wanted to do. I went to a U.S. National Team Championships in West Yellowstone, and there was a physiologist named Jeanne Wadsworth. She had done a little bit of work with Dr. Jim Stray-Gunderson and the U.S. National cross-country ski team. I thought these guys were really smart; they were using a scientific approach to training. I would just go out and hammer myself and train; maybe there was really a structured way to do it to get a lot better. I heard a lecture by Jeannie and was impressed, so I talked to Jim Stray-Gunderson a bit and ended up getting excited to do research and study cross-country skiing. I looked at a couple of different schools all located around snow country and found one in Michigan. After my first graduate course in exercise physiology, I thought that the key was training intensity. If you could just train at the right intensity you would induce the right kind of adaptations. If you trained too hard or too easy you would miss it. So I did my master's thesis looking at anaerobic thresholds in cross-country skiers. That research did not work out so well, so I ended up switching over and doing an anaerobic threshold study in the laboratory using cycle ergometry. What I wanted to know was if this magic training zone, the threshold training zone, would change or not. What I was really into was what the ideal training intensity for endurance athletes was, and how do you structure training to get athletes to spend a lot more time at those intensities.

Sean Richardson (SR): As an athlete, before you got into your studies, what was your take on training?

DM: I just tried to beat anyone I would train with. I wanted to make sure I was faster than the rest of the people. So I tried to get in with really good ski groups, and we would all go out and do 50 kilometers. Everybody would keep skiing along, and I would keep skiing and hope I was not the last guy who got dropped. It seemed to work pretty well; I trained hard and raced hard. I didn't hate anyone; I just got really competitive and wanted to see if I could do better than the other guys on the team. I didn't train with heart rates, and I didn't even monitor my training. We just had some coaches who said we're going to do this loop or these intervals, and we took whatever they dished out.

It sounds like Dave experienced, firsthand, what it is like to live the more-is-better philosophy. What Sean found fascinating in this interview was the way that Dave integrated these early athletic experiences into his work as an exercise physiologist.

TRAINING AS AN EXERCISE PHYSIOLOGIST

SR: So how did your training and your graduate work intersect?

DM: We started to do this research looking at the anaerobic threshold, and I started to see that when you study the physiology of a maximal steady state intensity, it gets very messy. It's not simple and clean and pure. There is no one ideal intensity; there are all kinds of intensities you need. Thresholds actually shift throughout training sessions. Athletes have threshold heart rates that after an hour are actually at different intensities than when they started. These magic training zones all fall apart if you are tired, or if you're fresh, or if it's hot, or if it's cool, or if you have wind in your face or not. So when I finished up my master's thesis, I thought there was no special training intensity. I started getting more and more interested in those periods of time where I had been really tired, and guys who I'd beat all the time would beat me in a race. I would go through these slumps; they might just be a bad race or two, but these nimrods who couldn't ski their way out of a paper bag were beating me. I didn't know what was happening to me. What is it that's lacking? Is my technique falling apart? How did I go so slowly in that race? I was becoming fascinated

> **" There is no one ideal intensity; there are all kinds of intensities you need. Thresholds actually shift throughout training sessions. "**

with the form changes, how they wax and wane. Sometimes at races I'd be yawning and barely dragging myself to the start of the race and thinking, "Why am I here today?" And then I would go and win the race and think, "Where did that come from?" So I became more interested in fatigue management. The big theme was overtraining, and I started to think I needed to learn a lot more about overtraining. I started to work with triathletes and cyclists predominantly, and during that time I started to see that genetically there are huge predispositions to be great in sport. But even those who are the most genetically gifted have problems with their form and have periods where they cannot perform well, and then they have periods where they've performed extremely well. The ability to predict performance is poor. When I asked people how could sport sciences help them, the answer was: "Make sure I don't get sick, make sure I don't get hurt, and make sure I will be in good form on the right day." Those were the big three, and I thought, "Okay, I wanna be the guy who travels around the world and tells everyone how to be in good form on the right day."

So I went to Wyoming to study endocrine markers of fatigue, like catecholamines, urinary cortisol, serum cortisol, salivary cortisol, and testosterone-to-cortisol ratios. Sport scientists thought that, in athletes, the endocrine system was going to unlock the secrets of when one has done too much and when one hasn't done enough. But after 5 years of research, I reviewed all my studies and had to conclude that there were no magic hormones; there were no magic catecholamines or testosterone ratios. The part that started to fascinate me most was how fatigue can be so specific. Some people can be tired for a certain performance task but not tired for others. People can be mentally destroyed and still pull off a personal best. Others can be mentally upbeat, or at least come off as highly motivated, and then get depressed because their performances are so bad. They think, "I am going to rip the lid off this next race, you watch me; I am feeling good," and then they are terrible. They try again—terrible again. To me, the uncoupling [from] the mental health model, where one needs to be this self-actualized, content person to perform well, was interesting. How athletes' perceptions, or at least what they tell you about their perceptions of what they can do, are uncoupled from what they can really do. There is a lot of guarded information in what athletes tell you; there are a lot of things they get really weird about.

> **" ...even those who are the most genetically gifted have problems with their form and have periods where they cannot perform well... "**

One can see that, early on, Dave was also interested in the psychology of training and what goes on inside athletes' heads. That interest in psychology has stayed with Dave throughout his professional career. Now that we know more about Dave and his history, let's move on to the parts of the conversation offering tales of overtraining from an exercise physiologist's perspective. Dave starts here with his current fascinations. In moving into stories about athletes, Dave illustrates his points about overtraining with his own tale of overdoing it, of sibling rivalry, of self-inflicted unhealthy training, and of confusion over shockingly poor performances and fatigue.

DM: I have started to become more and more interested in taking a performance task, train an athlete really hard at that task, and see how compromised that performance task becomes. Then the question becomes: How long do you keep that compromised state? And then with a taper, how do you unleash the load and watch everything bounce back? How do you train so you produce a really resilient system? Over the last 6 years I have been much more interested in tapers and keen to try and identify whether an athlete is responsive or not responsive to intense training and taper. Ideally, you load them, and they become tired and suppressed, and when you unload them they bounce back, and there is buoyancy in the system. I was just looking at some data this morning where we gave female cyclists a hypoxic challenge in a tent to simulate altitude. As you go from normal oxygen to low oxygen when you are tired, you can't do anything; you just lay there. When you give athletes who are fresh the hypoxic challenge, they sense it. Their ventilation kicks in and guards against the problem. To me, now, the ability of an athlete to deal with a dynamic challenge is where my interests lie. I have been real fortunate to travel, study, live with, and be around elite Australian cyclists where I can study this stuff.

THE SOURCE OF FASCINATION: CONNECTING DAVE'S STORY WITH HIS ATHLETES

SR: You seem to have a varied background and a long history in sport yourself, as well as many years working with athletes. You have also been working in the area of overtraining and fatigue management. Can you tell me any stories about particular athletes you have seen who have gone through that experience? What are they like?

DM: Let me start with my own story. One time when I was a sophomore, skiing competitively, I was having a really good year. I had won every race. Particularly, in the middle of year, I seemed to be in uncanny form. For me, it was the first time I had a couple of years of training behind me, and I was the top dog on the pile. Then we came to the end of the season; I had a chance to win the regional championships. I probably mispaced the whole event; I was probably not in very good form. It wasn't like my skis were slow or anything. I had lost all of the efficiencies, that floating, easy, fresh feeling of just flying over the snow and feeling really supple, having everything get into a nice rhythm and having fun with the race. It turned into a nightmare. I went out too hard. I had splits that were way up on all the top guys I was racing against and died a miserable death. I pulled off second, but I had the worst last five kilometers of a race that I've ever had in my life. I was really damaged from it, and Nationals were just 3 weeks later. I had an absolutely shocking Nationals where I couldn't get out of my own way. I couldn't figure out what exactly had happened. In retrospect, I look back and I had really, really great form for a long period of time, but I did not manage it well. I used it, and showed it, and played with it. It was so much fun to have form, to show everybody how fast I was. I would go out on training sessions and drop everybody. When everybody would go in after practice, I would go and ski some more because I felt great. I began to add in morning sessions before breakfast.

SR: You were almost getting too high on your own achievement?

DM: Yeah, I was. And now I have seen that happen with me training on my bicycle. Whenever I had good form on my bicycle and I could keep up with my brother (his brother is a high-level competitive cyclist) on the hills, or when I am starting to feel really competitive with him, when he is a little off, and I am really on, then I would push myself through a week and then become absolutely trashed. Then the following week I would find myself in a big fatigue hole. In the past, I used to get worried when athletes got really tired, but now what I get most worried about is when someone has really good form. I have seen it over and over in cyclists and skiers where they are having big leaps; they are leaner and faster than usual. They are fitter than usual; everything is going really, really well. These apparently positive changes are some of the key warning signals for me now. I used to just say, "Wow! He is in really good shape. Let's see if we can motivate this athlete to get his lactates higher. Let's start to really play with this new ability to overload you." To me it is crystal clear that when you see an athlete launching into newfound form, and is loving it, and is euphoric, that is the time when training management becomes most critical.

SR: And there is lots of positive feedback; you feel good about yourself.

DM: And it feels great, doesn't it? And a good coach, an experienced coach with a lot of grey hair will catch somebody in about a week of that, or even 3 days of that, and they hold them up. I have seen a female cyclist in the lead-up to the Olympics here in Australia. We were monitoring her power outputs over fixed-time durations, and 2 weeks out from the Olympics they were showing the best form we had ever seen in her life, and she had a long training history. So she had been around; she has been an Olympian, and a National champion. The motivation and pressure on Australian athletes leading up to the Sydney Games were unbelievable. Everything was so incredible, supportive, positive, goal driven. Things just kept coming together, and if you're on track it's like, "Wow, it's even going better!" She was just busting out of her skin. She was laying down amazing power outputs. We had to calibrate and recalibrate and check the crank because we could not believe it. Even with all the knowledge we had, we had an extremely difficult time trying to contain her because she was in this superwoman mentality. I reckon that when she hit the road race, she was coming off of feeling great because she had abused herself for 2 weeks showing us how wonderful her form was. At the Olympic time trial, which was a couple days after her road race, she had the most disappointing results we could imagine, probably one of her worst results ever at international competition for this woman, and it was at the Olympic time trial. She probably had a chance to medal or at least go top five. If we could have had her 2 weeks earlier for that time trial, she would have ripped the pedals off it.

SR: What would you have done differently, looking back?

DM: Probably would have been more conservative and come up with more ways to interrupt the loading pattern. If there is going to be a hard day and it is super, super brilliant, string them out for 4 days before they are allowed to do another hard day. Don't go easy day, hard day.

SR: How did you deal with her, psychologically? Here she is, super keen; she is on top of her form, and she wants to do more.

DM: Yeah, we talked about locking her bike up. We talked to her every morning, every night, every training session. We said, "Give this one a nudge and see what you have in you," and we would all just be floored. And we would say, "Okay, instead of doing four, we are only going to do three intervals up this hill, and on these next two do them, but don't bash yourself. Do them and leave the intervals wanting more, and do this well within your capabilities because a lot of times it seems that the 95% effort can be done over and over again, but it is the 100% effort that rattles you. Just hold back a bit, even though it feels good, hold back!" But she couldn't contain her enthusiasm. And we were subtle, but obviously not convincing; you never know how much to back it off. And it is so much fun to see it; the coaching staff was all marveling at her form. I saw it happen to a male cyclist who was an Australian national time trial champion. He was at an altitude camp in the best form we have ever seen over a 4-year block. He had some of the best form leading into the World Championships; he was just brilliant. He was launching into a new level, and he could not believe how good it was for him. We had an experienced coach and good support staff, and we all watched the form

> **" ...a lot of times it seems that the 95% effort can be done over and over again, but it is the 100% effort that rattles you. Just hold back a bit, even though it feels good, hold back!"**

show itself day after day, and we just marveled at how wonderful it was. He just fell into a heap after that altitude camp; he could not get out of his own way. He had a mediocre performance at the World Championships. It is very difficult to know how to use the best form you have. I think what happens is as the form starts to go, you see this real elusive goal slipping through your fingers. It is not gone, so you can't say, "I have lost it." It is just slowly being lost, so you think, "I will just try a little harder; it's not that far away." You never say, "Back off, it's going. Let it go, and it will come back." Instead, you keep chasing it to death, and that is when you run into the hole. So it just slips through your fingers, but it's not really gone. It is going, and we are going to let it go because we know that in time it's going to come right back. It is like, "I have never felt this good in my life, and I am just scared it's going to run away, and I will never see it again. So I will keep grabbing after it." Like this guy I'm talking about. He was going to the World Championships. His form had been great, and he just started to lose it, a couple of bad training sessions and he thinks, "Ahh, crap! I am getting close to the World Championships; I'd better try harder." Then you get in a stressed mode and start doing really weird things. This guy started going after supplements: "Maybe I need some creatine phosphate. Maybe I need bicarbonate. Maybe I need to do some carbohydrate loading. Maybe I need a little ginseng." Athletes start getting really desperate.

Another guy, again at altitude training [now a professional rider with an international team] was in the best form of his life at the camp. He was just ripping the pedals off. Everyone was saying every night, "You were just unbelievable." He was loving life; everything was going so well for him. Then after that camp, he went over to Europe, and he was in the biggest fatigue hole for 3 weeks. It was just terrible, terrible fatigue that he was dealing with and having a difficult time racing. But in all of these scenarios, the female Olympic cyclists, the two males who got really tired, all of them had come back to form. I have only seen one scenario that was particularly worrisome, and it was a female cyclist who was highly talented, and she had been a triathlete as well as a road cyclist. She was also a physiotherapist. She had a lot of demands placed on her by her job, and she was really Type A, meticulous, organized, always on time, a focused person. And she went through some good form, and then she got tired, and she was forced to do some more racing. After that racing, she was in an extreme fatigued state. She had performance incompetence; she started working really long hours, and she just got pissed off. She was so angry at her form, so angry at herself, that she started triple-timing her training. It was like she was really angry at herself, and she kept smashing and smashing, and she went into a state where it was the longest performance incompetence I have ever seen in cycling. This is only one person over 6 years in Australia working with the national men's and women's and mountain bike teams. She was the one person I saw actually go into a state where there were 2 months of just working and training really hard, coming out of it, and performing even worse. We spent a lot of time with her, the team went overseas, and she was performing so poorly she could not travel with them. She was depressed and sad about the whole thing, so we started a multidisciplinary approach with her. She was talking to a psychologist. She was seeing a nutritionist; she was working with us on really structured interval-power outputs that were important to be able to compete, but not much else, and lots of rest in between to try and coerce the body and mind back into saying, "Don't be too scared, you can come back." Every time we started to see a little bit of her old self come up, we would say, "We have licked this one! You are coming good; the volume and the intensity are coming better. The diet looks good; you are starting to mellow out a little bit with the job and the other stressors in your life." Then she would raise the bar. She would get one big training day in and fall into a heap for the next two or three. She could come back and show us just for a moment how good she once was, but she had no stamina, no ability to handle the volumes of intensity that she was able to manage before. She ended up quitting the sport. She couldn't come back for a whole year. Now she

> **" It was like she was really angry at herself, and she kept smashing and smashing, and she went into a state where it was the longest performance incompetence I have ever seen in cycling. "**

has come back, and she has done the Ironman in Hawaii; she has come back to be a really top-notch triathlete. To me, that was an interesting one because I have talked to other athletes, and there is certainly a pretty important issue, and that is that it is not just stress that comes about with training, but it is the other life stressors combined with the training. When people come to the Institute, their meals are made, their beds are made, their laundry is done; they eat, sleep, and train, and there is nothing else to do. When you have a job you are trying to hold down, kids you are trying to take care of, stressful situations with a relationship, or moving house, it is those things make it all a lot more problematic. I think they can really compound the stresses that go on.

Dave and Sean's conversation then moved on to coaches' roles in training and overtraining. Dave has a keen appreciation of expertise and a healthy respect for the knowledge and the struggles of coaches working with athletes who may be at risk of overtraining.

COACHES AND OVERTRAINING

SR: What can a coach do to assess the individual? You do all those physiological measures, but what about all those other things in their lives? How do you make those differentiations between individuals' lives and what the risk factors are?

DM: You start coaching, and then you start looking at mistakes, and you keep making notes of all the mistakes, and you look at everything that has gone well, and that is what a good coach is, a person who very rapidly assimilates all the observations that are in place before him. He says to himself, "I have seen guys like this before, and I have seen them come up on their form like you are right now, and I have seen big problems following it. I want to sit you down and talk to you about people you will know, people very much like you, personalitywise, backgroundwise, technique and talent–wise, and I will show you three of them who have all gone through what you are doing right now and have ended in massive crashes. I don't care if you crash after World Championships, but I do not want you crashing before. You are going to make the ultimate call, but I am your coach, and it is time for me to sit you down and lay out some of the observations I have made over 20 years to tell you why I am giving you this advice." Coach knowledge cannot be captured on a form, and it is not a POMS [Profile of Mood States] profile. It is not a blood test. Good coaches have great awareness and have great experiences; they don't get written down. They are full of historical experiences and anecdotes. Poor coaches can get way off track. Some unknowledgeable coaches think that her performance was because she rode a white bicycle because every person on a white bike has had a good race, so we should get white bikes. It has nothing to do with white bikes; it just so happens that over the last few years the top bikes had particular stems to drop people low, and they were white. This point is where sport scientists can help put resolution on the observations. For example, a coach may think that every time we go to altitude people get sick, but what is probably making them sick is the airplane flight because every time you go to altitude you have a big international flight following a big race—that is how everyone gets to camp. Going to altitude is probably not making them sick; it is probably going on international flights after 2 weeks of stage races that is making them sick. So why don't we try not getting scared of altitude. Let's try giving them 3-day windows after races at the competition venue for relaxing, lying by the pool, going for walks, watching some movies, and hanging out there instead of forcing them onto a plane. Good coaches, they do it on their own; that is what makes a good coach. The great coaches are the ones who are able to take a group of different individuals and, based on all of their experiences, roll with the punches, and go, "A little bit of this for you, and a little bit of that for you." If you ask them how, some of them may not even know how they do it; they just sense that their athletes are stressed and need time off.

SR: My question now is—for example, say you were going to a coaching conference, and you have a group of junior or developmental coaches, even novice coaches, and someone says, "David, we know about recovery and fatigue problems and overtraining, but what would you say are some of the risk factors? What should we be looking out for?"

DM: I would say, "It is a very difficult area. I cannot provide you with a cookbook explanation, but what I would suggest is that you become an intern, that you serve your apprenticeship with a master. You will pick up more doing that than probably from any lecture that I can give you. What I can do is go over the terminology so that you can talk the talk and read the papers. If we can get the terms established, then you can move quicker." What is probably one of Australia's real secrets is that they have scholarship coaching programs. The master coaches have apprentices with them every year who go out in the field and watch and observe and take notes. There are physiologists and support staff around as well, but the big thing they are doing is before big competitions. They sit there and watch the athletes and make their own calls. They cannot say anything—it is not their team—and then they see the master coach, and watch him, and see how he is handling them. "Why did you pull him out of that training session?" "I cannot talk to you about that right now, but I will tell you later. I just got a hunch." And they still see where the master coach makes mistakes, because they still all make mistakes. I usually see two types of coaches. One is just smash them until they break, and the other is this real nurturing, more personable type of an approach. Coaches who are really secure with themselves and have had great athletes before have got a lot of rapport with the athletes, and they can get into those interpersonal relationships. They are secure with themselves, they are happy with their coaching jobs; they have real good perspectives on the whole scene. They seem to say, "So, John, how are you? I remember when I was like you. I would really like to see you do great. I don't care if you have another coach or not; that is great. We are here for you to get better. I am not going to claim your success; it's yours. I am here to provide the best environment for you that I can. Let's see how we can work together." The coaches that are not as secure with themselves say, "By God, this is the way we do it. This is my program; it creates champions. If you can't cut it, you were never meant to be a champion. Don't talk back to me. I don't want to hear about your problems. I don't have time for that." It is a very autocratic, unidirectional, "put up or shut up" mentality.

Dave then talked about an area that makes many of us sport scientists uncomfortable. Although Dave has a great deal of respect for expertise, he has little tolerance for professional hubris. His candid stories below show that he too had fallen into the "I am an expert" trap.

STORIES OF PROFESSIONAL ELITISM

DM: I think that as sport physiologists or psychologists we have limited perceptions of reality. We get locked up in our labs; we get locked up in our universities. We read our papers; we look at our means and standard deviations, and we come up with a working model. Then we are comfortable with that model and our terminologies, so we talk to coaches in the format we are comfortable with. It makes us sound good: "You are in my lab; I'll talk to you that way. You don't know those terms, but I do. I'm great; you're not. Let me see if I can bring it down to your level." We say these kinds of self-important things, but in reality we're all scared shitless about being out in the field, because we're dorks out there. We're these nerdy scientists who get in the way. I just tell coaches that the best experience you can ever get is not going to come from a lecture. It's going to come from getting out there with a top-level coach and watching her do her stuff. Watch her deal with a superfatigued athlete. Watch her deal with a very fresh athlete. Watch her take an athlete who is superdepressed, but physically capable of doing some training blocks. Watch somebody who has an athlete who is mentally motivated beyond belief but who's got a body that's just failing on him. Watch how she deals with people who are outliers in the group, and watch how she deals with the group. That'll learn ya.

SR: Your point about being the expert is a fascinating one. Can you tell me some more about your experiences as an expert exercise physiologist?

DM: It is like, you've got your PhD done, you've published four or five papers, all in the area of risk factors associated with persistent fatigue in athletes, and you're really fancying yourself as a bit of an expert. Then somebody has some money with a professional team, like hockey, and they say they want to hire you as a consultant. "So, how are we going to monitor risk factors in this particular group?" You know, even with all your experience, and even if you're really stable, you feel, or I felt, compelled to have the answers. I felt like, "Dr. Martin, your time has come. Make the call. How are you going to help us here?" And the real answer is, "I don't know; I need more time with your group." Someone else might make a better call than me, but the ball is in my court. I've got all my tools. I've got financial support. So they want me to make the call on how to prevent overtraining and how to prevent persistent fatigue in these athletes, and the real answer for me is that I don't know how to do it. I say, "I've got some ideas I'd like to run with, but it is going to have to evolve over time, and we may go down a lot of avenues that don't become fruitful, but by me being around, I'll try to be a supportive person for you to make sure that you can do this a bit better. I am your adviser; I'll try to advise you appropriately. I really will defer to your expertise." That's the professional approach, but it's not always implemented. A lot of times they ask you, "So how are we going to prevent overtraining in this group?" They look at you very defensively, Mr. Expert, Mr. Know-It-All, Mr. PhD. "Okay, NASA Boy, let's see you do your thing." And I tried to, when I first came out to Australia. I was like, "Okay, well, I'll show you what we can do. I'll help you prevent these problems. We can make a better program. Follow my lead; we can do it." And I ran into huge problems. And then when the coach got disappointed with me, I said to myself, "Well at least the athletes will recognize my brilliance." Then you start chumming up to the athletes, and you form all these athlete relationships, and then the coach is getting even more disappointed. The athletes are caught in this web of confusion and never know who to trust, and it takes you 4 or 5 years to mature to the point where you are in a real stable relationship. I've got a job that's paying me all right, happy with who I am. I'm happy with my sport allocation; coach has got his job. He's on a 4-year contract; he's one of the most successful coaches we've got, and he's stable. I'm stable. We can look long-term over 4 years and see what's required for success and what type of people are going to help foster that progress, and we can each tell each other, "I don't know." [Shared ignorance is a wonderful and rapport-building thing.] The coach can say, "I don't know," and we can sit down with an athlete and say, "What do you think? You're tired; you haven't done well for 4 weeks. What do you think?" And we can talk it out, and say, "We've seen this happen before with that girl 4 years ago. You know her; she was the best person ever to come out of X [rural town] where you're from, and she had the same problems you are going through right now. You want to hear about it?" And they go, "Yeah," and you go, "Okay." But most sports never get to the point where the coach is stable, the support coaching staff is stable, and the exercise science staff is stable, and long-term relationships build up with the athletes. But many National team models are where everyone gets thrown together, big camps; everything is competitive for coaches and athletes. "Do what you can. If there's a problem, it's your problem. If there's success, it's our success." You see why it's mayhem?

Sean brought the conversation back to the topic of athletes and the root causes of overtraining behavior. Sean is a psychologist, and he couldn't help asking the sorts of questions that follow. Even though Dave is a physiologist, he appreciates the psychological and personal-history side of things. In these excerpts, Dave sounds more like a good sport psychologist than an exercise physiologist.

EXERCISE PHYSIOLOGIST AS LAY PSYCHOLOGIST

SR: Going back to the one cyclist who really went into a persistent fatigue. What's the mentality behind that? How does someone develop into that kind of person, instead of a person who realizes what's going on, and takes the break, or does what's necessary, or finds help? What's going on at the personal level?

DM: It would probably take a really good sport psychologist over many, many sessions to get to the point where you think you are really seeing her reality. It's really out of my area, but it seems like everybody's identity gets tied with something, and with athletes their identities are their sports. Some people have really robust identities—they're a boyfriend, they're an artist, they're an architect, and they're an athlete—but some people have very narrow identities, and sport is everything to them. For some elite athletes the whole thing is, "I want to be known as a great person. I want to be proud of who I am. I have chosen to be an athlete. I am not a farmer. I'm not a student. I'm an athlete, and I want to be the very, very best that I can possibly be. Every time somebody beats me that means I am not that great. Every time I win that means I'm great." So in time their sense of identity starts to fail, but they do not want to give up on being somebody who's worth being liked, or worth being loved, or worth being great and saying, "I used to

be great and now I'm nothing," instead of saying, "I'm a great person and I used to be able to really go fast." There's one woman cyclist who's a real joy to work with. She's a lawyer, and she's a concert pianist, and she's a good rock climber, and she was a state-level water polo player. She's one of these uncanny talented people, very bright, loves life, and lives it richly. If performance starts to falter, which it does at times for her, she likes to sit down and go, "I'd be interested in your perspective on why my form's down right now, because I've been thinking it through a bit, and I reckon it's the way that I've been real jumbled with living in a new location, going to sea level, going to 5,000 feet, going to higher altitude, coming down, and I'm starting to think that the interplay of the altitude, with never enough time to fully adapt, is just kind of nuts. I'd be interested in your thoughts, because I really have poor form now, and I wonder how I might manage it this time." And that, to me, is so real. There is also the other extreme. It may be a fear of failure; I don't know what it is, but there is a bit of almost panic about it. It's like it's not somebody running a race to see how fast they can go. It's like someone running away from a lion, and they're going to die if they get caught. It's that kind of difference. One person is going, "How fast can I go? Wow, that was cool!" and the other person is going "No, no, I'm gonna die! If I ever get caught, I will die." It's like they cannot see a life as a loser; they cannot see any way out of that mode. So, it's a seriousness about things for them that is really, really tough to break out of. To be fair, there're some pretty twisted people to deal with, and if I helped make them that way, then I'm twisted. They might get out at 11 o'clock at night and do that one-hour training run to make sure that their volume is high through the week. Some of that behavior has allowed them to become great, but that same behavior can go over the top and be the source of their downfall, and that's the funny thing. The very trait that makes you great can be the trait that destroys you in so many areas of life.

> **"It's like someone running away from a lion, and they're going to die if they get caught."**

SR: I'm trying to understand what might make some athletes more susceptible to this sort of thing, and we're looking at personal variables as well as situational variables surrounding the athlete.

DM: Some have the need-to-please disease; they just absolutely want the coach to love them and appreciate them so much that they'll kill themselves. You know maybe it was with their dads or something like that. Maybe they never felt like their dads loved them, and now they've got this coach, and the coach gives them only a little morsel of respect and praise if they win a race, but that's so reinforcing. There have definitely been very abusive relationships with female cyclists and their coaches. I know of one female cyclist who has been physically, and what I feel has been psychologically, abused. I have heard the story that the coach of this woman yelled at her in public and said, "You stupid bitch! You know how much money you cost us by not winning that race?" That's a bizarre, twisted relationship, but that woman has raced very well on many occasions. She's come out and done very well at races, but it comes from a really demented, weird relationship. It's like, What's motivating you to be great? If you're being motivated to be great for healthy reasons, hey, we don't know what the healthy reasons are, you will be able to fall back in crisis times and take care of yourself and then come back again. If you're motivated for pathological reasons, you often can't break out of them. For some reason it has locked you into a perspective, and you can't get out.

BURNT COOKIES

DM: I have one story or analogy that has worked well with me when discussing fatigue. I've tried a lot of different ways to talk about fatigue with the athletes because I don't want them to become too fatigued, but I don't want them to be scared of fatigue either. So I came up with this analogy. See what you think. It's a baking cookies story, and the analogy is that you say that if you're a baker and you want to bake really good cookies then what you're going to have to do is put a bunch of ingredients together. You make up your dough, and then what you are going to do is put these cookies, or let's say *a cookie,* on a sheet and put it in the oven. You are going to bake it, and as you bake it a whole bunch of things start going on in the dough with the sugar and the baking powder and everything else. There are a lot of interac-

tions, and there're changes, chemical and physical changes. The chocolate in the cookie melts, and things happen, and everything heats up, and then you pull the cookie out of the oven, and the cookie cools. Then there's a certain time when you want to eat it. It's just warm and really good; it's not so hot that it burns your tongue, and it's not too cold but still a little bit crispy. And I say that the analogy here is to go through and think about the process of baking. You've got to have good ingredients. You put them in a bowl, you put the little cookie on a tin, and you place it in the oven. You've got to bake that cookie at a certain temperature, for a certain duration, and then you've got to pull the cookie out. You want to eat it at just the right time so that you get the full enjoyment. So then I go to the athlete analogy. The ingredients are like genetic makeup and background history. If you don't have good ingredients you'll never make a great cookie, so if you don't have good genetics you'll never be a world champion athlete; that's a given. So now you take that athlete, I mean the cookie, and you put it in the oven. The athlete you put in training. Training is the stimulus to promote these physical and chemical changes and adaptation. There're a lot of ways to bake a cookie. You can go low heat for a long time, or you can go for a high heat and a short time, and they bake differently. There're different ways to bake, but you can still get a good cookie with different ways of baking it. Some people have high-volume programs, and other people have high-intensity programs. There are different ways to become great. When you go through the baking process, you pull the cookie out of the oven, and it's just like pulling the athlete out after the overload period. You don't want to eat the cookie then;

> **" you can always go back and recook an uncooked cookie, but you can't uncook a burnt cookie. "**

it's too hot, and it will burn your tongue. You don't want to compete right after you've loaded because you're too tired, but with time the cookie cools, but if it cools too long, it's just cold and stale and too crispy. You want it when it's just right, and with the taper, you want to allow fatigue to dissipate, but you want to perform, and really capture that moment when your fitness is high and your fatigue is still good. If you go too long, your fitness will deteriorate. You can't take a cookie out of the oven that's too hot, bite it, and say, "Oh, you're a bad baker. You've cooked it wrong; that's bad cooking." What I'm saying is you tried to eat it too close to coming out of the oven, just like an athlete going through a great training program and trying to compete within a day or two after the overload. It may have been a great process with magnificent overloading, good variability, and good specificity. It's been excellent, but they are trying to compete at a time when they're fatigued. It doesn't mean the training is wrong; it means that they've mismanaged the fatigue. They should have waited longer, or done the loading earlier before the competition. Overtraining is when you take the good material, the good batter, the good dough, you put the cookie in the oven and you burn it. If you burn the cookie, it doesn't matter when you try to eat it; it's burnt. You've burnt it, and even if it's appropriately cooled, it's still burnt. When you are loading yourself, don't burn yourself and don't undercook yourself, otherwise you're just doughy, and if you undercook yourself, it doesn't matter how long it cools, it is just not a cookie.

SR: But the funny thing is it's a really good analogy because if you undercook yourself, an undercooked cookie is still all right.

DM: Yeah, it's fun to eat; it's just not as good as it could be.

SR: But a burnt cookie is terrible. Undercooked athletes are still all right; they don't feel terrible. They just don't perform as well as they could.

DM: And you can always go back and recook an uncooked cookie, but you can't uncook a burnt cookie.

SR: And it can depend on the day. Some days you put it in the oven for the same amount of time, and it just doesn't come out all right.

DM: And some ovens are different. You've got conductive air, and you've got different top heating, and you never have exactly the same ingredients. You've got two athletes, and you reckon they're just the same. So you cook 'em the same amount of time. Then one's spread all over the pan, and the other's hanging right in there. Yeah, and it just shows the complexity of it all. The good bakers, they do it over and over again, and they start to go, "Yeah, with this kind of oven and with these kinds of ingredients, this is how we want our cooking to progress, and this is when we want to eat the cookie; that's the scenario." What I want to do, when I talk to people about overtraining, is hear about burnt cookies. How have you burnt the cookie? Don't tell me about eating cookies when they're hot, don't tell me about eating cookies when they're cold, don't tell me about cookies that were perfectly cooked. Tell me about burnt cookies, and then they're like, "What do you mean burnt?" I say black; I want to see black, burnt, fried cookies. That's where the hard part comes is when you're talking about fatigue in athletes, because everyone has had a cookie even when it's a little too warm, or a little too cold, or a little underdone, but burnt? Once you get the terminology right, everyone's seen burnt, and they then say, "Oh, you mean *burnt.*" I then usually say, "Okay, let me tell you about how I burnt some cookies, and to me that's going on with some of your athletes. I hope we learn a lot about fatigue management issues, but the ones that really stand out are burnt, really fried athletes. Now what's led up to that scenario?"

At the end of the conversation with Dave, Sean was left feeling enlightened and hopeful. As an athlete, Dave had been relentless in his go-in-hard approach to training, completely absorbed in the more-is-better dogma. Yet he has transitioned to an insightful, psychologically minded physiologist who looks for balanced ways to the get the best performances out of athletes. Dave seems to think holistically about the athletes with whom he works. He sees a person in front of him with goals, motivations, and blind spots, not just a machine producing a bunch of heart rate graphs and power outputs. Sean ended the interview thinking that Australian sport is lucky to have a guy like Dave looking out for its athletes. Mark's interview with Dave would be sure to shed more light on the intricacies of working with athletes and the issues of overtraining and injury.

FURTHER CONVERSATION WITH DAVID MARTIN

Mark interviewed Dave in early 2006. The two share a relatively long history. They both immigrated to Australia in 1994 and had met each other at the University of Wyoming in 1991–1992, when Mark was a visiting professor and Dave was a doctoral student. Over the last 12 years, they have had numerous discussions about physiology, psychology, and what it means to be in service to others. In the following conversation, the focus is on the practice of exercise physiology and the relationships between physiologists and the people in their care—the athletes and the coaches. Before they get to service, Mark and Dave revisit the topic of inter- and intrapersonal factors that lead to maladaptive behavior in the realm of overtraining.

WHY DOES OVERTRAINING KEEP HAPPENING?

Mark Andersen (MA): We see the phenomenon of athletes, time and again, overdoing it with their training. Why does this repetition keep happening?

DM: It happens all the time, and you know, I don't know why. I think it's out of frustration for a lot of people. They train, they perform, they train, they perform, they do better, they train more, and then they perform and do not so great. Then they train more and really get worse. Now that's incongruent with the desired scenario: Train, get better. How can I train and get worse? So I'll just train harder, and I'll get better. But with an injury, throw that in, and it's a little bit of a funky sideline. "I train and my performance is better. I train more and my performance is even better, but my wrist is the size of a basketball. I train, and I've got tears coming out of my eyes, and my wrist is like a huge balloon. I'm still holding even, and there's this paradigm in my head that if I train hard I will get better. Now the wrist is just a problem if I can just stomach the pain; when that inflammation subsides I will get better. If I can just handle this training block into my taper, everything will heal up, and I'll be fine." Now why do they keep going? I suppose if I had to make a guess, I think most people when they start off never ever really know the limitations of the human body. You don't know with drinking what your alcohol limitations are. You don't know the problems of pulling all-nighters for exams.

You just don't know limitations until somebody comes along and says, "Hey, I've seen this before. I'd back off. The best thing you can do right now is rest." But who are *they?* They're not you; they've never seen anyone like you because you're special. This guy's just an advisor (or a new coach), and he's gone off his history, but he's never seen anyone like you, so he doesn't really believe in you. You believe in yourself, so you're just going

> **" Train, get better. How can I train and get worse? So I'll just train harder, and I'll get better. "**

to keep on going. You'll prove everyone wrong, and you will, and you're completely convinced you can overcome the odds, probably just like someone who's been diagnosed with cancer: "I'm not going to die." It's not until you really meet your own frailties and your own limitations that you're bottomed out. Then you go, "Oh my God, I've reached my limits. I'm going to learn from this; I don't ever want to get here again." But some athletes don't learn the lessons and [referring to the Bill Murray movie wherein a character experiences the same day over and over] are in a *Groundhog Day* state.

They are reliving a nasty experience. You can see it, and you wonder why they can't. I think the reason they relive it is because they don't see it as the same problem. They say, "Last time I did this, I didn't have anti-inflammatories. This time I do." "Last time I did this, I wasn't giving myself 5 or 6 hours between the training walks. This time I will." I think there must be some strange rationalizations going on. If they saw this repetition problem clearly, they'd be really hard-pressed to justify their behavior of doing it over and over again. I think a lot of them come up with some type of excuse. It's hard to understand why athletes would abuse themselves in the same way they did before unless there are other things, some behavioral and emotional tendencies that are so strongly ingrained that they can't escape them. It's like you have this response, and that response is not good for you, but it's what you do.

MA: Yeah, there's about four or five themes in psychotherapy that come up over and over again. One of them is, "I'm so anxious about something happening that I will behave in ways that ensure what I most fear will happen, does happen."

DM: That's perfect. These overdoing athletes are exceptionally motivated people. But what is it that fuels that self-defeating fire? Why are you so incredibly motivated to risk life and limb, and try to put the blinkers on, and not deal with the blood coming out of your ears, or the inflammation happening in your shoulder, or the dismal performances, or all the advice coming your way? Somehow you're resisting all of that and you continue to persevere. Probably the closest analogy I have outside of sport is little boys having temper tantrums. [Dave and his wife have two boys.] They start to act in ways that are not going to get them what they want, and it's just going to end up with them getting even less and less of what they want. They escalate, and then it blows up to a full tantrum.

MA: And they're lost in it, and they can't get out of it.

DM: And in the world of sports, there aren't many multiple winners. So your expectation is to be the world's best. If that's what you're trying to achieve, you're going to be a bit disappointed. How do you work with athletes who are on the road to pretty exciting things, but they start to move into these behavioral patterns and these downward spirals, whether it's chronic inflammation, whether it's some kind of a stress fracture, whether it's some kind of really persistent fatigue, and they keep sneaking out to engage in unneeded and damaging training practices? The coach is saying, "You don't have to." The support staff is saying, "You don't have to." How do you work with that? I really think that the only way to move them through any kind of modification of behavior is to intervene during some calm points in the athletes' lives. You've got to have a really strong and sincere relationship with the athlete.

> **" It's not until you really meet your own frailties and your own limitations that you're bottomed out. "**

THE EXERCISE PHYSIOLOGIST–ATHLETE RELATIONSHIP

MA: That's where I wanted to go next and hear you talk about the exercise physiologist–athlete relationship. I've been at the AIS, and I see people in service to athletes who cover the range from cold fish to Dave Martin, who they all love. What things actually fuel change? If you want to change an athlete's behavior you can't change it by being Mister Physiologist.

DM: You can't hit it with an analytical approach. I'll talk on behalf of physiologists. Exercise physiologists are trained to try and understand how fit an athlete is and how they respond to training stress. So we come from a real analytical point of view. What can I measure about you that tells me how fit you are, and how can I monitor those traits over time to understand what kind of training works for you, and so forth? Just about any young sport scientist can do that, but when you see something going awry, I think the best advice early on is to be really aware of everyone around that athlete, and try to understand who has the best relationship with him or her.

Is it the coach? Is it the sport psychologist? Who around the team is most (and healthily) connected to the athlete? When I started at the AIS, Clark Perry (sport psychologist) had tremendous rapport with the athletes. If I saw something going on that was funny, or something that was alarming or weird, I'm probably not the only one who's seen it, and I would talk with Clark. But if I'm going to act on it, sometimes it's not best to act on it through the athlete because I don't know the athlete. It may not be best to act on it through the coach. I've got to be aware of who's got the strongest relationship with that athlete, and then feed them with what I'm seeing, and then it's up to them to figure out a path. So early on I worked through people who had the best relationships, such as Clark. How can you have a good relationship with anybody in a year?

I reckon within a year, even if you're phenomenal with your skills at interacting with people, that you're probably just starting to get at the tip of the iceberg. I don't think you can be, within a year, the one who's going to have the rapport to deal with these real and serious issues we're talking about here. We're not talking about saying, "Yeah, it looks like you are really on good form." Those things are quite easy, but if you want to get to someone who's basically in self-denial or someone who's gone down a real problematic behavioral pathway, you're not even going to try and be the one to broach the issue. Search out and find the people around the athlete and be part of the team that's going to help, and let the spokesperson be the one who has the best relationship with that athlete.

Mark then nudged the conversation in the direction of negative examples of service delivery and how those in service can make things worse.

WHEN SERVICE GOES PEAR-SHAPED

MA: Now, do you have any stories about exercise physiologists who may have tried to take on that role of "handling the problem" prematurely, and have actually had a negative influence.

DM: Okay, I recall a bright young exercise physiologist, nailed it in all his classes, came with the highest recommendations. He gets allocated to a couple of sports, and one of the sports is going to be volleyball. He takes it on with passion. He wants to be exceptional, and so he starts to look at how the programs are designed and developed, and he's not happy with the low aggressiveness, and the "fuzzy warmth," and the desire to win quickly, and he believes that from a conditioning point of view there's a lot of things lacking. He comes on board, and he's not going to just assess things. He's going to intervene and start talking about how to train. He's not only going to talk about how to train, but how to implement these physical characteristics into the strategy that's going to take place in the games. He got to the point where he sat down with me and said, "What do you do when you're working with a coach who's a dickhead?" And I thought, "Well, this is probably not a good start." What's going to happen when a service provider is so self-assured and cocky and zealous? What's going to happen is, as athletes start to find themselves with problems, there is this person coming into the interaction with a big chip on his shoulder, and there's a whole bunch of things that are going to go on. One is that a lot of blame is going to get cast. "The reason you're in this state is because your coach is a dickhead, so there's a big problem. The second thing is I've got a training program, and if you're not responding to it, it's because you are not doing it with intent." The young physiologist comes in with all the best intentions and is frustrated with the way he's been dealt with. He's not been treated as a major player and does not feel supported by staff. He's going to start seeing every problem as either the coach's problem or an athlete's problem. The interaction is going to be really threatening; it's going to be confrontational, and it's going to be an interaction that's not going to change anything, really. It's going to make everybody upset. So there's one example, but there are lots of them. I knew a physiologist with triathlon, trying to make a big name for himself, and he tried to intervene with a lot of high-intensity interval training programs. The athletes were going into races very fatigued, and

then he would blame it on the psychologists and the other people in charge of recovery. "Those guys are screwing up your recovery. Nothing is going wrong; you're fine. I'm fine. The program's fine. Your fitness is fine, but these people are not dealing with your nutrition, your psychology, and your recovery, and so you're going into races tired, and, if they get off their backsides and start doing the job as good as I do mine, you'd be fine." Everyone breaks down and gets injured, and the whole team starts getting upset. So, whose fault is that? "Well, that's the soft tissue therapist again, and that's the fault of the physios because . . ."

MA: The relationship is why it's breaking down, because the relationships are bad. These are one-way unidirectional relationships, and they're exploitative. "I am here in order to make a name for myself through you."

DM: Exactly, exactly.

MA: And I'm invested in this stuff big-time. So if anything goes wrong, it can't be me; it has to be you, or it has to be the coach, or it has to be somebody else. That's enough horror stories. Tell us a good one about relationships.

In the following section, we get a glimpse of Dave and his interpersonal relationships with the athletes in his care. His points about time are good ones for neophyte sport scientists to take to heart. It takes time to be effective, and time is needed to establish a caring, supportive relationship that fuels change for the better.

THE QUALITY OF RELATIONSHIPS FUELS CHANGE

DM: About 6 years after I came to Australia, I was down in Melbourne, and one of the athletes I worked with early on in my time in Australia came by, and we were having coffee, and we just saw each other, and it's like you've seen an old friend, and that was a really fun relationship for me, because I was learning a lot about myself as that relationship developed. She was a young talented cyclist, and I was just starting, so I was trying to figure out what was making me tick, and she was trying to figure out what was making her tick. It could've gone pear-shaped [gone wrong] a hundred different times, but it didn't. She never won a medal at the World

Championships, but she went through a lot of soul-searching and very difficult times. She was the young "chosen one," and then she had to deal with growing up and maturing and becoming a woman with other talented girls coming into the sport. She had recently missed out on selection. I had been with her through skinfolds and looking at how her body was changing, and realizing that she was maturing and growing into a woman, but she was feeling like she was fat, and I was wondering, "What does this mean?" I guess the thing that I was most proud about in that relationship was that early on I respected people who were more senior to me because I liked them, and they supported me in my neophyte status. I was more than happy to be introduced to these athletes and work with people who would say, "It's all right to be frustrated, David; it doesn't always work out." There were a lot of things I was trying to get a handle on at that time. My good relationship with the athlete coincided with a point where she was going through really tough times. After 6 years, you feel like you've been in the environment, you've built up the trust, and you have respect so you can say, "I'm the one who's got to tell you 'X.'" I'm looking around at the coach who's new. I'm looking at all the other support staff, who are new. I said something like, "I know we can talk about this stuff, and I really want what's best for you, and if you don't like what I am saying then you can tell me to fuck off. But you will also know I care, and I want the best for you." If a hard call needs to be made about backing off of training, the athlete knows I am not saying, "Forget the World Championships." Instead I am saying, "You need a break, and let's talk it over, and you're a smart person. I've been around you for 6 years. Let's talk together and identify a strategy to get on top of this problem. You are tired, but we're not giving up because then you just detrain. You are not going to stop riding because that's important to you, but we're maybe going to remove you from some competitions, and we're going to change the way you train, and we're going to change your role on the team, and maybe if you like that, let's run with it, and if you don't like it, let's come up with two or three things that you'd be happy to work with."

> " ...the important thing is that we both, together, come up with a solution we buy into, and then I'm not going to run out the door and leave you. "

You pick which one of those you want to do, and it's not about me being the boss. It's about knowing each other over 6 years. I think we're ready to start talking about some solutions. If it goes a couple of ways, I'll tell you I'm not happy with it, but the important thing is that we both, together, come up with a solution we buy into, and then I'm not going to run out the door and leave you. I'm here to walk through this stuff with you to see how we keep moving forward."

MA: It's the relationship.

DM: Yeah, and I mean that's probably a really important part of when you interact with the athletes early on. If you're around for 2 years, well that's pretty impressive. Hardly anyone works with an athlete for more than a year. If you do 2 years, you get to the point where you are saying, "I am a resource for you. I'm not trying to make you dependent on me. I'm a sounding board for you, but I cannot fix you. I may help you get to where you need to go quicker, or I may just be a person who observes some counterproductive behaviors over and over again, but you know what? I'll still be here." One of my really deep-seated desires is not to be loved by the athlete but to understand how athletes engage in training and performance and be there when they have extraordinary things happen to them and when they have really disappointing and dismal things happen. So that's kind of my little message to the athlete. I have to do a lot of soul-searching too. What am I getting out of all of this?

This question sparks Mark to start grilling Dave (in a collegial way) about his motivations for being in the profession of exercise physiology. There are few better questions for us professionals in service delivery to ask ourselves than: Why are we doing what we do? What are we getting out of our profession? How are we fulfilling our own needs?

WHY DO I DO WHAT I DO?

MA: What *are* you getting out of it?

DM: Well, I'm learning. That's what I think.

MA: And what are you learning about?

DM: I'm learning about individuals who are struggling with the high goals that they've set up for themselves.

MA: But what are you learning about yourself?

DM: I'm learning about how I am trying to get it right. I'm learning about how I deal with life from how these guys deal with life, so there's a little spin-off there, but I also . . . [Pause]

MA: Sorry, not meaning to be such a snoop.

DM: No, it's good, and it's interesting. I've thought about it a lot. I probably get a couple spin-offs. What I think I get out of it is: I'm in a profession where I am employed to try and assist talented people in achieving their sporting goals; that's my profession. So, the more I can learn about people who have done it well, and have done it poorly, the more I can try to help create referrals, or teams, or approaches, or implement some types of relationships, or some types of communication, or some types of awareness. The more I do that then the better I can be at my job. I also don't see any difference between service provision and science.

MA: That strikes a major chord.

DM: "Are you providing a service, or is this research?" We've got to always categorize. Actually, it's both, and that's what I'm saying. Without a doubt, my life is one big long scientific, and service, journey. I am trying to observe, quantify, evaluate, refine, hypothesis-test, and serve. It's one big long continuum, but if I'm just a service provider, I'm not a researcher, I'm not a scientist. I reckon that split is just pissing in the wind. I think you're wasting a lot of time because you're not carefully making observations. You're not trying to get to the root of the problem. You have no big constraints or frameworks or models in your mind that you can use to piece all those observations together to create understanding. Every girl who burns herself out, and never trains again, and gives up the sport, as bad as it may have been for her, if I've got a good relationship with her, and I understand her well, then I've learned heaps, and,

hopefully, the next time I see someone struggling like she did, I maybe will be more competent to help. I've been around for 13 years in this business. Pretty soon I'll start seeing the kids from the girls I've worked with. It's been a fascinating journey.

MA: Yeah, in psychology we also spend a lot of time looking at the instrument of service, the practitioner, him- or herself, because service doesn't occur in a vacuum. It occurs in a relational context, and Dave is the instrument of exercise physiology service.

DM: Yes, and if you don't understand the instrument and implement what it knows, you are not as effective as you could be, and the correlate to that is: If you cannot implement, and you cannot convey your observations, and your thoughts, and your intuitions, to someone who can help, then you're not just ineffective; you are possibly dangerous. Well, that may be overstating it, but maybe not. You have to know the people who can implement and call them in.

MA: Aaron Beck called psychotherapy "collaborative empiricism," meaning that two people were involved in a scientific kind of endeavor, trying to figure out the experiential world of one of them though exploration, hypothesis testing, measuring, thinking, and trying new things. It seems that description fits well with your view of the role of the exercise physiologist.

DM: It's perfect.

MA: Thanks, Dave. We'll talk more . . . as we do.

REFLECTIONS ON DAVID

David is a great storyteller, and his skill as a raconteur is one of the reasons we chose him for this chapter. His storytelling helps put a human, and messy, face on the world of athletes and training. Many of the tales he told boldly illustrate the risk factors outlined in the previous chapter (e.g., risks from foreclosed athletic identity, abusive coaches, psychopathology, being in great and "invulnerable" form, and pushing it over the edge). David's stories of overtraining and service bring to life the complex, messy, convoluted, and tangled real world of athletes, coaches, and sport scientists struggling with trying to achieve excellence in the best ways they know how. What comes through most clearly in this chapter, however, is David's humanity. He is an expert, he is a scientist, and most of all he is a human being. He has a solid understanding of and empathy for human frailty (his own and that of others). He knows the power of human connection. He knows that all the bells and whistles and heart rates and lactate levels won't really amount to much unless there is a healthy and caring relationship between the athlete and the service provider.

I (Mark) am in the fortunate position of being in continuing, albeit sporadic, contact with David Martin. His encyclopedic knowledge of exercise physiology, and his keen appreciation of the psychology surrounding the relationships between sport scientists and the people they serve, have been a bedrock for me in my work with athletes in Australia. Dave is my Platonic ideal of an exercise physiologist. He would blush at my accolades, but that is Dave. We need more people like Dave if we want to understand what it is we do when we encounter the people we were trained to serve. As a psychologist, I spend a lot of time in the realm of emotions and cognitions. My conversations with Dave help ground me in the body.

CHAPTER 6

SPORT SYSTEMS CAN DAMAGE

Conversations With a Sport Psychologist

When we were in the planning stages of this book, and decided we would have two chapters on individual professionals, we immediately thought of Trisha Leahy. Her approach to sport and overtraining (OT) is both challenging and thought provoking. One of this book's core themes is that of taking a wide and holistic perspective on the problems of overtraining in sport. Trisha takes this theme to its logical end, prompting us to focus on the whole sociocultural system of sport and how it is one, if not the major, driver of overtraining.

INTRODUCING DR. TRISHA LEAHY

When one first meets Dr. Trisha Leahy, the accent is hard to place. She was born in Ireland but has spent a great deal of time in Australia and Hong Kong (she speaks fluent Cantonese). She has been practicing as a psychologist for almost 2 decades. For the majority of that time, as well as running a private practice, she has been employed at sport institutes (the Australian Institute of Sport and the Hong Kong Sports Institute). She has extensive experience in working directly with client populations who happen to be elite athletes, in the context of Chinese culture in Hong Kong and in the context of Australian culture. Her doctoral dissertation concerned trauma and sexual abuse in sport. Her research has been on the cutting (and taboo) edge

of applied issues in sport psychology practice. Her work was deemed so important that she was asked to present a keynote speech at the International Society of Sport Psychology's quadrennial conference in 2001 in Greece.

Trisha began her tertiary education at the National University of Ireland, where she received her BA in psychology. In 1990 she completed an MA at the University of Hong Kong in comparative Asian studies on the Filipina domestic labor force in Hong Kong, and in 1995 she received her Master of Philosophy in psychology at the Chinese University of Hong Kong. She completed her PhD at the University of Southern Queensland in Australia in 2001.

Her current position is head of Athlete and Scientific Services at the Hong Kong Sports Institute (HKSI). She directs a team of more than 60 professional and scientific staff members who provide support for the elite training programs at HKSI. She oversees services and research in sports medicine, psychology, biochemistry, exercise physiology, biomechanics, and nutrition, along with athlete career, education, and social development programs.

Sean conducted two interviews with Trisha 5 years apart. The first took place in 2001, when Trisha was employed as a senior psychologist at the Australian Institute of Sport (AIS) in Canberra. The second was in 2006, after Trisha assumed her current position at HKSI. Please be warned: Trisha is confronting and controversial, and that is why we wanted her voice in this book.

OVERTRAINING, ABUSE, AND TRAUMA

Sean Richardson (SR): I hope today we can have a discussion on the sorts of things that you have come up against in your profession working in sport, where you may have seen or experienced athletes who were pushing too hard.

Trisha Leahy (TL): The issue of this *pushing too hard* is actually quite interesting to me. My other area of interest is in trauma, and I see these two issues as integrally connected. When I say trauma, I mean perhaps childhood trauma, which could involve things like violence, family emotional abuse, physical abuse, and sexual abuse. Traumatic experiences can happen at any point in someone's life, and they can result in serious psychological sequelae [aftereffects]. Some of that impact, or some of that trauma, can get organized around overdoing things, whether it is in sport or getting to the top of the corporate ladder. So it's not something that's specific to sport, in that sense. In terms of pushing it too far, I have rarely seen issues around overtraining that are not associated with some quite significant underlying psychological issues. I don't want to give the impression that we can think about overtraining in a simple linear fashion. Because A has happened, then B, the outcome of overtraining, is going to happen. If we think in that simplistic linear way then we are going to miss the complexity and all of the latent variables that might be going on around someone's experience. I think elite sport is particularly facilitative of this type of pushing-too-hard behavior in that it generally gets rewarded. So you train hard; you're always doing that *extra mile* kind of thing, and you get rewarded. So it legitimizes what, in another context, we may look on as perhaps self-medicating or arousal modulation behaviors, such as drug and alcohol abuse or eating disorders. People use various methods to modulate their anxieties and distress. Overtraining is a nicely rewarded and sanctioned way of doing that. I am speaking in hugely generalized terms, of course, but these connections are some of what I am seeing.

> **" We in sport, it seems to me, have come to accept a high level of abusive behaviors that we would not, for example, accept in the classroom. "**

SR: Can you tell me a specific story that comes to mind?

TL: I've worked with many clients who have experienced severe abuse growing up. They then come into a sporting situation where we have a sociocultural environment that in many aspects replicates the dynamics of abuse. You have a hierarchal system where enormous power is vested in coaches, and they are oftentimes only accountable for their athletes' performances. We in sport, it seems to me, have come to accept a high level of abusive behaviors that we would not, for example, accept in the classroom. This is problematic, because it replicates the abuse dynamic; you have a powerful person who cannot be challenged. Because you are an elite athlete, you are seen as something different, and this can be isolating in itself. Other people don't interfere with those kinds of coaching behaviors because it's elite sport and therefore perceived to be special and different. We replicate the dynamic of a dysfunctional family, where one person has all the power and kids are being abused. No one will dare interfere with the family because we assume the family is good; we assume sport is good, so we don't interfere. In this way, we open the sports environment wide up for abuse at all levels, including sexual abuse.

I've just done research looking at these issues with athletes, and what I found was that where athletes had experienced emotional and sexual abuse, they were extremely distressed on psychological measures. But emotional abuse by itself, without any physical abuse, which I also measured, or without any sexual abuse, was itself strongly predictive of psychological distress. In sport we have normalized, I feel, an emotionally abusive environment, and we have just accepted it. There are two problems with this situation. One is that it replicates the abuse dynamic. The other is that it masks other types of abusive behavior such as sexual abuse, because emotional manipulation and abuse are the primary weapons of a sexual perpetrator. So to me all of these issues are what's going on with how people organize themselves around coping with being in an abusive situation. If you're an athlete and the coach is this powerful person, and you are away from your family, then that coach can become the primary role model, the primary attachment figure in your life. Attachment is a psychobiological need that we all have, so to maintain attachment, you will seek approval of the primary attachment figure, and in sport, approval will be there for going the extra mile. Now if the primary attachment figure is emotionally volatile or arbitrarily gives out reward and punishment, then you are going to have, from an attachment theory perspective, a very disorganized attachment.

SR: It's security? Or confused attempts at gaining security?

TL: Yes, exactly! Disorganized attachments. So, why are we surprised when this athlete is trying to do everything perfectly and overstretching every single way so that they can have some emotional safety? To me those are significant issues that we need to examine.

SR: It's fantastic to talk to you, because I think you come from a different perspective, and it really makes a lot of sense. I think that these areas aren't what coaches and doctors and physiologists usually touch on. Sometimes we tend to believe that athletes are these great people, and they come from these great families, and they don't have these issues.

TL: Well, that's the mistake we make. We mistake highly functioning in one task-oriented area of their lives for how they are functioning in all other areas. Of course, they are not highly functioning across the board, and the problem is not just with athletes. We make these assumptions about everyone, such as actors, CEOs, presidents of the United States, who are quite successful in their jobs, but often have messy personal lives.

SR: And I guess it's an area that a lot of people certainly don't want to deal with.

TL: No, they don't.

SR: Too touchy or sensitive or scary?

TL: Yeah, it is, and these issues are what my PhD has been about. We don't know how to recognize it because we don't understand these dynamics. Coaches are not taught about this kind of dynamic, and so people see athletes who are smiling and happy because they are trying to

maintain attachment to this key figure in their lives. The athletes look happy and are smiling, and people think, "Oh, but they're happy." When there is trauma involved, or what we call "traumatic attachment," then there is a type of disorganized attachment. To use an extreme example, it's like what we typically see with hostage situations, where the hostage is taken captive, totally deprived, and is emotionally and physically abused. Eventually the victim becomes dependent on the perpetrators, who constitute the only allowed relationship that the hostage has, and who are the only providers of basic physical and emotional needs. The hostage starts to attach, and that's what's called traumatic attachment to the perpetrator. We see this all the time in people who have been held in captivity for a long time, people who have grown up in very abusive families. You get this traumatic attachment. You can get a similar thing happening in sport, and that was one of the themes that also came out of my research and also from my experience working as a clinician with traumatic attachment and trying to therapeutically unravel those issues for athletes.

ARE YOU UNCOMFORTABLE NOW?

Trisha's take on sport environments, coach–athlete relationships, and abuse is confronting and makes many of us in the profession uncomfortable. Some sport psychologists might protest that she is blowing abuse in sport out of proportion, and maybe in some ways she is, but that does not negate her arguments. The response, "Oh, she's over the top here," may be a kind of defensive posturing designed to dismiss the points she is making. We need critics and iconoclasts like Trisha to blast away at the shibboleths we have in sport. That said, despite Sean's stated enthusiasm, he does at this point become somewhat uncomfortable with the topic. He tries to move the conversation to something less emotionally charged.

SR: You have worked in Hong Kong for quite a while, and you've worked here in Australia. Do you think that in terms of the pushing-too-hard and overtraining mentality there are cultural differences?

TL: Hong Kong's an unusual place because at a sociocultural level sport is not valued as much as in other countries like Australia, and there is not yet a deep sport history. There are not many indigenous or local role models that would encourage young people to participate in sport. Also, a proportion of the people in Hong Kong would have come originally from China, during politically unstable times. Many of these people would have been coming as refugees or escaping from the cultural revolution, or the 1940s revolution. They are often seriously concerned with basic and material survival and want their children to be doctors and lawyers and successful business people. Sport is certainly not seen by them as a legitimate career path. So that's a different environment than in Australia, which is a sport-mad country with lots of role models. Coaches in Hong Kong would sometimes complain to me that they were having problems the opposite of pushing too hard. They couldn't get athletes to train seriously enough. They just wouldn't, so you, kind of, had the opposite end of the scale happening over there in some instances.

In mainland China there is a different system. If you can become a successful athlete, you not only can become more well-off; you may get rewards from the government, and your family benefits also. If you are living in a small flat or house, and you've got 10 other relatives living there, and you're out in the middle of the country-side, and you happen to be very tall, and you get picked up for a basketball team or something, then you may become the savior of your family. So the potential for significant overtraining is there. Additionally, the cultural and sociopolitical discourse of sacrificing individual well-being for the good of the motherland is very compelling. . . . So you can have a lot of casualties, but there are always many more talented athletes in line waiting to fill the gap.

SR: A war of attrition?

TL: Exactly! It's the opposite in Hong Kong and different from here [in Australia] in that it's much more compelled by, perhaps, poverty and the need to take care of your family and bring "glory to the motherland." So there are quite different systems at work.

SR: Looking at it from a developmental perspective, have you ever had the experience where you see some of these cultural issues come out when someone's been removed from their culture? For example, when mainland China athletes move to Hong Kong, or even when they move to Australia or somewhere else.

TL: Yeah, in terms of that translation from one culture to another, for the young people coming from China, there was a bit of a myth that, "Oh, Hong Kong is part of China, and it's the same." It is in many ways similar, but there are significant differences also. And so there can be dislocation and alienation. Even though athletes in Hong Kong could speak the same language, the way of thinking, of conceptualizing the world, is rather different from the mainland Chinese young people. They would be extremely obedient to the coach and extremely hardworking, and then you'd have some of the Hong Kong athletes who were like, "Yeah well, I've got to go now, because I have to study." The good thing is that this cross-pollination of influences can be, and has been, very positive for both groups.

These cultural difference issues will come up again in the second interview, conducted 5 years later. For now, Sean directs the conversation back to the original topic: examining the risk factors for overtraining. He also prefaces his question in such a way as to further avoid the issues of childhood trauma.

EXPLORING PERCEIVED RISKS

SR: Within sport do you think there are different situations where you might say that athletes don't have that overdoing-it mentality? Maybe they haven't experienced major trauma as children, yet are they at risk in certain situations? What sorts of environments or situations can put that person at risk for injury or pushing too hard?

TL: Apart from the trauma issue, the way that sport replicates abusive hierarchical power and powerlessness dynamics is a risk for everyone who's involved in that system, whether one has a history of trauma or not. If you want to be successful, you have to survive the system, and if the system itself is abusive, then you have to adapt to an abusive system, and dysfunctional behavior is one of the rewarded things. Even without trauma in the background, overtraining can happen. I think it is also fueled by the media and the whole hero-worship thing that we do around athletes, especially in Australia. It seems that nearly everyone colludes in this myth of "no pain, no gain," and you're a hero if your leg is broken and you can go out there and win the game for your team. This ethos is extraordinarily abusive from my point of view, where we all collude, and the media colludes, and the athlete gets called a hero.

SR: Why do you think this collusion happens?

TL: I have no full explanation; part of it is the dysfunction of sport, and part, I guess, is our dysfunction as a society. It's related to why do we go and watch movies where people get killed and assaulted, and we call that entertainment as well, so then why would we expect anything different on the sports field? It's entertainment.

SR: Do we need this abuse to have elite sport, or can there be an elite sport with a modified attitude? I wonder if elite sport is always going to be like that because it's part of the definition of elite sport, or . . .

TL: Well, I think that's what needs to change. Of course, it has been tried, and there are coaches and athletes who don't operate within this abuse dynamic and also do very well. The truth of the matter, however, is that punishment, negative reinforcement [e.g., training extrahard to avoid the coach's displeasure], and abusive behavior *do* produce world champions. Negative reinforcement and abusive behavior have been methods of social and political control since the beginning of time. Of course, such a system is successful; it achieves the same goal, so that's not the question. But is that what you need to be elite? The question is: What are the costs of such a system? Would you buy into that system, and do you want your child in such a system?

SR: Right, so it's a question of: Are there other alternatives?

TL: Yes, of course there are. We have enough good coaches and athletes who don't live like that, who also achieve very well. Part of what needs to change is the culture that we have in sport. Here at the AIS, I have been involved in the development of the Harassment Free Sports Strategy policies that are going to be implemented for all sports and clubs, and they're addressing precisely these issues.

> **" Does that child have an option to actually say to the coach, "I feel really humiliated and hurt when you say I'm fat; I'd rather you didn't." The child does not have that option... "**

SR: You mean at levels all the way down, Australia-wide. [There are now policies in place and workshops on those policies being delivered across Australia.]

TL: Yep, exactly.

SR: I think that would be critical. I've been involved in coaching at school levels, and I've seen the psychological abuse from head coaches, and the kids just accept it.

TL: Exactly.

SR: For example, coaches, in front of other athletes, talk about an overweight player. You get coaches who'll say, "Well, he's a fatty, and he's got to lose 20 pounds," and the kid himself laughs about it, and goes, "Yeah, I'm a fatty." That laughter has got to be masking a lot of hurt.

TL: What options does a child have in our culture with the way we accept the power of the coach? Does that child have an option to actually say to the coach, "I feel really humiliated and hurt when you say I'm fat; I'd rather you didn't." The child does not have that option, so the only option is "ha ha" and pretend it's a joke. That's how we deal with harassment. That's how we deal with situations that are threatening, and we're trying to get out saving as much face as possible. So this issue is problematic, and it's extremely intrusive of personal boundaries. Can you imagine what would happen if teachers in the classroom spoke to kids like that?

SR: Parents would be up in arms.

TL: Precisely! See, this is what I don't get. Why do we accept that abuse in the educational area of sport, and we don't accept it in the other educational area of the classroom? It is unbelievable to me that we accept this contradiction. What are we teaching children? The children don't have any choice. They don't have access to an alternative viewpoint where someone is saying this behavior is actually not okay. In the situation where parents don't question the coach yelling at their children, what is the message to the child? "People are allowed to abuse you and invade your boundaries, and no one is going to do anything about it." Why are we surprised when the child ends up with all sorts of dysfunctional coping strategies like overdoing it. What we are effectively saying is, "If you feel hurt and sad about this, it's your problem, kid."

SR: Canadian culture is not very different from Australian culture, but there was enough of a difference for me to question, "Is this the way it is here?" As an assistant coach here, I didn't do that sort of thing, but it is hard for me to step in and say to the coach, "Hey, this might not be so great for these kids," because that would be undermining his role (and threatening to my position as an assistant).

TL: That's exactly right, and that's what keeps the system going. That power structure where it's clear to you, as well as to the child, that no one is allowed to question the coach. This situation is what we need to change in Australian sport. I don't, however, think it's unique to Australia. In the UK, from colleagues I've talked to there, they say the same thing; it's a big problem.

Sean has been trying to move Trisha down the path of looking at individuals and individual risk factors, but he has not been very successful. Trisha keeps coming back to the system and the culture. Her position is coming from viewpoints more in keeping with social psychology and sociology. Sean wants to get down to the individual, the personality, but Trisha keeps bringing the conversation back to deeply rooted systemic problems. Sean makes another attempt to bring the conversation back to the personal risk factors, but he is not terribly successful. Trisha's point is that if we don't manage the big picture, then anything we do at the individual level is still going to have to operate within the larger dysfunctional framework.

THE SYSTEM IS THE PROBLEM

SR: Someone shows you a group of athletes, and they're new to a program. What would you say would be things to look out for? Like risk factors, if this athlete is at risk for going overboard, for getting injured, for wanting to please too much. What would you see as both the personal, and then

maybe some of the situational, environmental, external factors?

TL: You are asking the question what are the risk factors in individuals. When we go down that track, we are essentially depoliticizing the whole issue of overtraining and identifying it as the problem of the individual, and not the problem of the system, and I think that's a mistake. I think nobody in sport should be at risk for overtraining. Sport should provide an environment that facilitates excellence in a positive, empowering, and compassionate way so that no one needs to overtrain to get rewarded and to get to feel fulfilled. Sport should be the place where past trauma gets resolved. Sport should be the place where you get a different view of yourself from your abusive childhood where you were a little shit, and you had no value, and you knew people didn't care about you. Sport should be the healthy place, and we idealize sport and believe it is such a place. But, the way it's structured right now is the opposite of our idealization. Everyone is not equal, and everyone does not get a fair go.

Of course, everyone is not equal in sport. Some people have more talent than others. Some will be chosen, and some will not. But that is not what we believe Trisha is saying. The point is that sport should be a healthy place, and not one where horrible lessons are learned and psychological traumas ("you are worthless") are repeated. Sport should be embracing, not excluding (and that includes the messages that exclusion conveys). People who are less talented should also be able to find places in sport where they can demonstrate their competence to the best of their abilities (and be praised). In Australia, a teenager can get a fair go in the Under 16s footy (Aussie Rules) B League, and if he doesn't make it there, then he can have another fair go in the Under 16s footy C League. Trisha's point is that everyone who wishes to should be able to find a "loving home" somewhere in sport.

INDIVIDUAL RISKS VERSUS SYSTEM RISKS: FALSE DICHOTOMIES

SR: That's a great point, but if we have to work within sport, can we say that some individuals might be, within that system, more at risk than others.

TL: Hmm . . .

SR: I feel you are uncomfortable with that question.

TL: Yes, I am. There is no point in asking questions only at the individual level because by doing that what we're saying is that we can only have people who can adapt to this extremely dysfunctional system. In other words, we want people who are used to dysfunction and who can function in dysfunction. I mean that this stance is an exploitative and unethical position to take.

SR: I'm thinking of the individual who might go into a sports situation where they have a coach who is understanding. They've got some supportive role models, and a coach who is trying to pay attention, and then they go out and train more on their own. They have this idea that, "I've got to do more." Now, I think that comes down to the sports system as a whole, maybe not their little microcosm in the sports system. Their coach and their team, at that moment, are good, but the sports system as a whole is creating that drive.

TL: Exactly, so again it comes back to: It's not the person with these risk factors; it's the system. So you're right even with the microcosm of that person's own team environment being positive, you still have the media, the hype, the enormous pressure in the wider system. Athletes, especially in an institution like the AIS, watch each others' coaches, hear what other athletes' coaches have said, and what other athletes have said. So, a young athlete may think, "Well, even if my coach is really nice to me and says everything is okay, but someone else, my friend or her coach, is saying, 'Oh yeah, you know you've got to really push out there, no pain, no gain,' you know, then already I'm getting anxious about my amount of training."

SR: Like, "Maybe I should be doing more."

TL: Again, it is the system at issue where you have inconsistency of approaches.

SR: Within sports what do you see as some of those situational risk factors, more at the sport-specific level, not so much a cultural level?

TL: You mean specific factors within a sport itself? Um, it's so hard to separate them because everything is just so multiply determined and has to be contextualized multiply as well. So you can't just pick out something. I feel, I could be wrong, but what does come to mind—and I absolutely admit that this observation is superficial thinking—is what I call *body-heavy* sports like rowing, triathlon, and cycling. They may attract some people who are working out some issues through extreme bodywork. Those sports are so heavy on the body, and that line between what is training and what is overtraining at the elite level becomes tricky. I think that's a bit of risk factor. The other risk factor that occurs to me is sports like gymnastics where children are being put into intensive training regimens at an extremely young age. This can be very problematic in cases where it is the parents who have the elite sports gold-medal dream for their children and push them much too early into intensive training. When parents are living their lives through their children, that's another area where you're possibly going to have something happening with overtraining.

PSYCHOSOCIAL AND FAMILY DYNAMICS

From this point, Sean moves the questions of overtraining to the possible influence of family dynamics. This shift in the topic, and his descriptive example of a reasonably functioning but highly motivating family, as well as its influences, sound suspiciously like Sean is describing his own family (see his story in Chapter 10) and asking Trisha to interpret part of his own experience. Sean is probably also describing the majority of families who have members involved in sport, so his question, although personal, is getting at many athletes' and families' experiences.

SR: Looking at families, it's probably more obvious what's driving that athlete in terms of family issues. You might be dealing with an athlete who is pushing really hard or has some major issues, and you try to find out what's going on at the family level, but it is not really clear. I'm not sure how to frame this. I can think of examples where you would have apparently quite a functional family, but it still engenders a lot of drive and motivation in a child, as opposed to a dysfunctional family where you have a real fear of removal of parental affection, or seeking approval because of the dysfunction. Do you have some examples?

TL: Yes, all the time, and this problem is something I often talk about with coaches as well, especially in relation to eating disorders. The factors around eating disorders can be also similar to the factors associated with being a successful elite athlete, such as conforming, perfectionism, goal-directed drive, and so forth. That's the thing, you can achieve your gold medal because it's the fun thing to do, and you know you are skilled at the sport, and you are talented, and you have a great family, and everything is really happy, and that motivates you to say, "Okay, let's see how far I can go." Or, you can be someone who has no self-worth because they've never been given any from day one. That person, who's also motivated and driven to achieve the same thing, but from a position of, "This will save me; this will mean I'm of some value," and, of course, it never means that at all. Once you've got your gold medal, then you have to figure out what's the next thing, because, "I still don't feel like I'm of value now. I've got to find something else to go and do. What am I going to do now?" I have oversimplified quite a bit in this illustration, but the point is that motivation can come from extremely opposite ends of the happiness and self-value spectrum, and it's a very complex issue.

SR: And it is really dependent on the context as to how that motivation gets manifested.

TL: Absolutely. I also have an example of an athlete who came to train in the elite system and was from this happy family. She came from an isolated country area, with not a lot of experience about the world, or about how things are. She was extremely talented in her sport and so got picked up into the elite training program. She had this expectation that sport would be like it was back in her little hometown: full of the joy of training and of everybody being happy. So she gets to an elite program, which is at that point being run by an emotionally abusive coach, with lots and lots of boundary intrusions, forceful physical contact—for example, holding someone so tightly by the arm that they're bruised—and lots and lots of emotional abuse around body issues and how people look. That young woman, coming from a stable background, ends up with eating disorder problems, overtraining problems, and is in extreme distress. So this scenario can happen to people with the healthiest backgrounds and points out how abuse affects everybody.

SR: Yeah, someone who came from a different background and had grown up with abusive

sport, like you say, might have developed a coping strategy to deal with it and been okay.

TL: Of course, yes exactly. Well, not been okay.

SR: Not okay, but being more functional than she was.

TL: Because they already have their good coping strategies in place, they are able to cope with it, and also, if I've grown up with that then that's normal for me, that's what I expect, so it's not surprising to me, whereas it was such a shock to her because she hadn't grown up with that abuse.

SR: It's interesting, because this idea of transitions and new challenges has come up in a few of my interviews: the situation where athletes are at risk when they move from the country, move away from their parents, and I think talking to you has really highlighted some of the reasons why there is a problem. You are going into an abusive situation when you weren't used to one, or if there's attachment problems being away from home and family. Transition itself is a stressful time, and it leaves people vulnerable.

TL: Yup, but again for me it comes back to: If the system that you step into is healthy, then your vulnerabilities are taken care of and you're safe, emotionally safe.

SR: So how do we change things? How do we safeguard the system?

HARASSMENT IN SPORT

TL: This problem is what we are working on with the harassment-free strategy. You should have a look at some of those documents. There is a whole series of booklets around keeping sport free from harassment. There are guidelines for sport and recreation clubs, guidelines for coaches, guidelines for athletes, guidelines for administrators, blueprints of how to set up a harassment-free policy in your club. There's a document on preventing homophobia; there's a document on disability. There's a child protection and sexual abuse document. So Australian sport is finally, in the year 2001, trying to get policies implemented and trying to sell these policies. [See Web page sources at the end of this chapter.] We have, at the moment, four sports who have agreed to act as kind of role models for everyone else. They're involved in an intensive year of working out the policy and getting education programs implemented for all members of their clubs, their national associations, and their executive committees. I've been involved with this initiative, and we meet once every few months to see how things are going and what input they need. We are trying to figure out how to structure it so that it will work out. Then we may have a good blueprint to be able to say, "Look, it's been done by these four sports; you can do it to."

SR: That sounds good.

TL: So we are working on it from the systemic level.

SR: It struck me that some of the things we were talking about, coaches may not be aware of these issues. I am hoping that we might help people become more aware. What do you think would be some of the things you'd like to say if you were to give advice to coaches, athletes, parents, and administrators?

TL: At this point the advice is, for coaches especially, be aware of yourself and the assumptions that you are making about how to relate to people, and young people particularly. Be aware of people's boundaries, and be respectful of those boundaries. For parents, it's about being aware that you can actually demand that another adult not abuse your children. It is not acceptable. And so parents need to be educated as well about these issues. Then, of course, we need to help athletes to be aware as well, but we can't leave it up to the athletes to change the system. They are relatively powerless to effect change, and they are dependent on the system itself. We can't just be passing the buck to the young athlete who has no power anyway to do anything about it. We need to make the change at the systemic level. So, it's coaches, it's administrators, it's managers, all system-privileged support staff who need to take responsibility for identifying abusive behavior and saying it is not acceptable. And there are ways of doing that where it doesn't have to be humiliating or confronting for the person who happens to be doing the intrusive or abusive behavior.

To me there's no point providing information where we say, "Oh look, these are the kinds of athletes who are vulnerable and at risk." Then some coaches may just say, "Oh yes, okay. Well, this athlete clearly has a problem. If they can't survive my system then that is too bad because it's the athlete's problem, not mine." We have to move away from that approach and say if we want change to happen, it has to happen at the top. [Silence.] So I think I take a very political approach to my research and practice.

SR: It's wonderful to get such a different perspective. Doing this interview-type project, I get so many different perspectives, and each person I talk to has a unique contribution. It really has opened my understanding of the processes that go so far beyond individual overtraining. I think what I'm really learning is that overtraining is just one manifestation of all kinds of other of underlying problems in sport. Like you say, it's a real systemic issue and is connected to other issues such as gender, masculinization in sport, and how that promotes that "tough" mentality.

> **"** *For parents, it's about being aware that you can actually demand that another adult not abuse your children. It is not acceptable.* **"**

TL: Absolutely right, and I think that's an important point to make that men are particularly harmed by that ethos.

SILENCING MEN'S VOICES IN SPORT

SR: But women, in a way, are trying to imitate that which has made men successful. For women at the elite level, it is still a very masculine approach but maybe not so much at the participatory level.

TL: No, you're quite right, but another of my "soapbox" areas is the way that we silence men's distress in our culture. We don't talk about the sexual abuse of men; we don't talk about the fact that a man might be emotionally in pain or in distress. We silence all that, and we have the unidimensional media heroes, who are able-bodied. They're white; they're heterosexual and they're powerful and in charge. And they can't be victimized. That's a very harmful myth.

At this point, Sean did not pursue the issue of how sport can damage men, probably because of his own discomfort with the topic. In the second interview, this topic will be examined in more depth.

SR: The topic of overtraining has just exploded into a whole universe for me.

TL: It's hard to look at these issues without taking both sociological and the psychological bodies of literature into account, because that whole masculinity issue is just so important and gender relations issues are crucial.

SR: The next stage of this project is going to be talking to athletes. I started off with criteria for choosing athletes, saying, "Oh, I want athletes who've been diagnosed with real identified issues with overtraining," but as I've gotten more and more into talking to people, I am realizing that the criteria are too tight, because it really applies to almost all athletes. I mean the ones that it doesn't apply to, I don't know where they are.

TL: Yeah, you're quite right. So you only get half the picture, or a small part of the picture, if you only go to where people have been diagnosed.

SR: Thanks so much, Trish, for taking your time to talk to me. You've left me with lots to think about. I am sure we'll talk to each other again.

TL: I am sure we will. That would be fun.

On that note, the interview wound down. Now we move on to 2006. Sean interviewed Trisha again, this time by a Melbourne–Hong Kong telephone connection. The following dialogue revisits topics from the first interview, with new twists, as well as other topics, such as the role of mental toughness and the problems that men face in sport. Sean and Trish begin by revisiting trauma and abuse, but this time the mechanisms that lead to overtraining are explored in more depth.

FIVE YEARS LATER IN HONG KONG: OVERTRAINING AND TRAUMA REVISITED

SR: We'd talked 5 years ago about this issue of pushing too hard and how trauma is connected to the issue of overtraining and stress–recovery balance. I might ask you, what's your perspective on that now, and do you have any stories about athletes who fit that perspective?

TL: Working from a biopsychosocial perspective, I think it's important to look at things holistically. I still think one of the factors that I'm interested in and being aware of is: What are the underlying drivers to behavior that prevent somebody from successfully achieving a stress–recovery balance. I have a lot of stories on that topic. One athlete I worked with was chronically going into overtraining and injury, then recovering from the injury, then going back out and repeating the pattern over and over again. When I see a pattern like that, I'm particularly alert. In a simplistic sense, learning isn't taking place. "Why do you keep doing this over and over again?" I don't mean that in a blaming way or in a pathological way. But, I have to ask, "Well what's going on here?" I'm just wondering whether, with this kind of persistent overtraining, if the repeat injuries are directly related to both childhood and sport experiences of psychological, sexual, or physical abuse.

SR: Without disclosing details of the athlete, could you give us a picture of perhaps what that means or what was going on there?

TL: The athlete I am thinking of came from a quite chaotic family, and the family itself was the perpetrator of the abuse. So the athlete is growing up in an environment where abuse is normalized, and therefore learns no coping skills, or no skills that other children will learn in terms of recognizing danger and being able to get out of dangerous situations and being able to protect oneself. Because the abuse started so early, none of those protective mechanisms were ever learned. What we know about children who've been severely abused from very early on is that there are high levels of dissociation. By dissociation, what we mean is a split in consciousness, memory, or identity. In other words, a split between the experiencing self and the observing self.

SR: And how does that lead into the injury and overtraining?

TL: The dissociation makes for a disconnected, or a fragmented, life story. So parts of life that have been intolerable get disconnected from the whole narrative. The dissociation operates in such a way as to allow the person to keep repeating similar patterns without recognizing that it's a pattern because the dissociation just segments each block into a discrete episode. Now, in that kind of situation there's a lack of awareness of self as a total being, and there's only awareness of fragmented parts of self. So it's very hard for learning, and making connections, and joining the dots together in a whole-life picture. The therapeutic level of intervention is extremely in-depth and long-term. Now, if that young person, who's already been harmed and hurt by the family environment, then comes into a sporting system that replicates, inadvertently or otherwise, aspects of that kind of disempowerment and denial of personal needs and personal entitlement, then you would—

SR: So sorry to interrupt, but how would you describe to the lay reader what you mean by "replicates the dynamics." If coaches were reading this interview, what would you say, or how would you describe to them what those dynamics are?

TL: In terms of replicating the dynamics, the core dynamic of an abused child or a child in an abusive family is that the child has no rights or entitlements. The child's needs are irrelevant. The only thing that's relevant is the needs of the adult. Now if we're in a coaching environment where we place similar emphasis, where we say the only needs and entitlements that we will take care of, or that we will acknowledge, are those of the coach, and as long as the coach gets us the gold medal, from the system's point of view those are our needs. However the coach gets that medal is not that important as long as we get the outcome that we need. So then accountability of process gets lost. So in that way, then, a disempowerment dynamic happens where all the power is vested in one authoritative person whom nobody, including the system and all the support staff, will question. The athlete feels or experiences him- or herself as very isolated within that closed environment. There's no access to help; there's no way out. The athlete feels trapped. Then dissociation has to happen to be able to cope with that, to be able to survive the two juxtaposed versions of reality, one of which is, "I'm on the path to the Olympics and a gold medal," and the other reality happening at the same time is, "I'm being abused by the same people who are helping me to achieve that gold medal."

> **" ...yes, this coach is extremely volatile and looks psychologically abusive, but this is elite sport... "**

This situation is totally disempowering. The same thing happens in an abusive family. The family sends the message of, "I love you, I love you," but at the same time is hurting the child. This is what the sports system sometimes does. It says, "I care about your well-being so I'm going to get you the gold medal," but at the same time it's allowing abuse to happen, too, to facilitate getting that gold medal.

SR: And then what sorts of behavior do you see in that athlete?

TL: Behaviors in the athlete come from an inability to modulate, or a lack of awareness about how to modulate, both emotional as well as physical recovery. Then a repetition of patterns occurs of always doing too much in an attempt to please those in authority and in an attempt to feel safe. With the athlete that I had in mind, it was a particularly severe case. I'm not saying that this abuse is the norm in all of sport, because it's not. But it does happen, and the system has allowed it to happen. This athlete was in a situation where the coach was not only extremely emotionally and psychologically abusive but was also sexually abusing the athlete.

SR: Oh, God!

TL: So you have an exact replication of the family abuse dynamic right there. It's really, really hard for an athlete to be able to survive that

SR: Oh, that's horrific.

DUTY OF CARE AND PROFESSIONAL BYSTANDING

Now that the mechanisms of dissociation, repetition of family dynamics, and the politics of abuse are a bit clearer, Trisha leads Sean into a professional practice issue that revolves around responsibility, duty of care, and what happens when we just let things go on as they "normally" do.

TL: Yeah, it is horrific, exactly. Now the question that I would raise, and that interests me, is: How did this happen in our sports system? With so many of the professional and support staff around, how was this allowed to go on? We who are the system-privileged staff have been, for too long, standing by and basically taking on the role of bystander and not challenging, not questioning, and not critically evaluating what we're seeing. We have been the ones who've been saying, "Well, yes, this coach is extremely volatile and looks psychologically abusive, but this is elite sport, and hey, it must be okay." Anyway, we don't dare question it because he's the head coach, or she's the head coach, and that person is so well-known and so famous. I experienced quite a bit of that through the research that I did when I was in Australia. What came out of the research was, in situations where athletes had been abused by their coaches, those who were more severely hurt or distressed in the long term were those where the coach had primarily used a methodology involving this kind of volatile emotional reward system. It's quite capricious and arbitrary. The athlete never really knew what was going to happen next because in one instance the coach would be extremely caring and nice, and in the next instance the coach would be ranting and raving. Now the bystander effect happens when the people are standing by watching this situation and not questioning it. One of the athletes told me her parents would stand by and watch this abuse, and they became at some point quite uncomfortable with the way the coach was behaving, but because the athlete herself was so enmeshed in that system, she was saying, "No, no it's okay; everyone is like this. This is just the way it is in elite sport." She was a very bright young teenager, 13 or 14. Her parents saw that none of the other coaches or support staff were questioning the coaching tactics. So the parents think, "My daughter is saying that it's normal; therefore, maybe that is elite sport. Maybe it is normal; maybe we should just trust her." Of course, what was going on was this coach was sexually abusing the athletes on the team and using emotional volatility as a way of keeping them psychologically off balance.

SR: You mentioned that you're talking about the extreme example of sexual abuse and that doesn't happen the majority of the time, but certainly emotional volatility is something that happens even without the sexual abuse, and it's still quite problematic.

TL: I agree, and I think this is one of the key sociocultural vulnerabilities and risk factors within some of our elite sport systems that will ultimately harm athletes and contribute to things like overtraining or long-term psychological distress. It's time for us bystanders, and especially those of us in the sports sciences area, to start becoming more proactive about demonstrating best practice and about contributing to coach and system education around best practice. We have taken steps, for example, in Australia and the UK at national policy level to promote best practice and to stop this kind of psychologically abusive behavior that we somehow have normed in the sporting environment. I think I said this last time as well: If we were to translate those behaviors into the classroom, there's no way that we would allow that kind of teaching and instruction to occur. The question is: Why are we allowing that in sport? We really need to critically evaluate that question. Why have we allowed these norms to continue? They're so harmful.

SR: I think one of the issues that came up last time was: We allow this abuse to happen in sport because abusive behavior does produce world champions.

TL: Of course it does, but if we put abuse and oppression in the political realm, we question it. We don't sanction it in our education systems; we don't sanction it in our families. We have laws against the abuse of young people. So why is it that in sport that we can't address this question? It's a question of ethics. Also it's a question of if you can get to be the best in the world and have your gold medal through the use of empowering environments, scientific training, and well-planned and joyful and empowering systems, then why not?

SR: Why are we having trouble selling people on that second idea? It sounds much nicer to me.

TL: Exactly! That's the question we need to critically evaluate from a systems level, and we need to look at where the barriers for that are. Partly, the barriers are ourselves, the bystanders. We haven't been standing up and promoting alternatives or mainstreaming alternative ways of achieving excellence in sport.

SR: I think sometimes it's actually just easier for coaches to push athletes really hard, and in being abusive they don't have to think much about what they're doing. To balance an athlete and get them just right, to stroke them and bring them along carefully, is so difficult. It takes more effort, I think.

TL: It does take more effort. I have another story. An athlete had come out of an extremely psychologically abusive coaching system with lots of bullying and volatile behavior on the part of the coach, and won a gold medal. I met that athlete in my office, and the athlete just threw the medal on the table and said, "You tell me how to value this; it's of no value for me because the whole process of getting there has just been just too horrible." As we worked together, gradually the story came out. This athlete was showing signs of post-traumatic stress disorder, and the question for me is: What are we doing in sport? We're traumatizing young people to get gold medals for us at the Olympics. Is that the price that we have to pay? I think we can do better than that, and I think that's not an acceptable way of being, but in that athlete's experience nobody, including the psychologists, including the medical support staff, including the other coaches, nobody explicitly questioned what was going on in a way that would have supported the athlete. Everybody saw it, and everybody apparently accepted it as, "Well that's the norm; we're getting the results, therefore . . ."

Sean, once again tries to steer Trisha away from what he considers the "extreme" end of the spectrum back to the more subtle forms of coaching and training practices. Trisha, once again, brings him back to the idea that these "extreme" forms of abuse are really not uncommon in sport, and she has the statistics to back up her claims.

SR: Yep, now we're talking about some of the more extreme cases of sexual and physical abuse. What I'm interested in getting a bit more perspective on from you is that there's going to be a lot of stuff in between great coaching and abuse. If I'm going to do a presentation to coaches, and I hit them with, "Okay, look people, athletes are getting abused, and this is what's causing the stress–recovery imbalances, and we've got to take care of athletes' injuries," they are going to turn off. They will probably think, "You're a bit far out there." And then they will just shut down completely and not pay attention to some of the more subtle things that people are doing. What do you see with respect to some of the more subtle, yet still damaging, behaviors that are going on?

TL: Okay, but first of all let me just say that as part of the research that I was doing at the time when I was in Australia around this more severe end of actual sexual abuse, the prevalence figures among the population that I studied were actually quite shocking.

SR: Okay, what sort of figures?

TL: Of the male athletes, 21% of them said that they had been sexually abused at some point before their 18th birthdays, which was my cutoff date for child sexual abuse, and over 30% of the female athletes said the same thing. Now those figures are actually quite comparable to other countries' data in terms of sexual abuse of young people. When I then further investigated those who had said they'd been sexually abused and asked them where the site of the abuse was, 41% of the abused women and 29% of the abused men reported that the abuse occurred within the sports environment. [Note: "Sexual abuse" here is defined as physical contact of a sexual nature, as distinct from "sexual harassment," as in verbal sexual suggestions and innuendos.]

SR: Of the ones who had been abused?

TL: Yes, now that's a very high rate, indicating there are some serious systemic issues that have to be looked at. To give you a comparative example, we could go through a few schools around the region and check how many young people had been sexually abused before they were 18. We would probably get similar results to the 20% for men and 30% for women, but then if we asked those abused young people how many of them had been sexually abused within the school environment, the figures would not be anywhere near the ones for sport environments. So those are extremely worrying figures, indicating that there's a systemic issue here. There's a vulnerability within

> **❝ ...the athlete just threw the medal on the table and said, 'You tell me how to value this; it's of no value for me because the whole process of getting there has just been too horrible.'❞**

the sports system that's allowing predators to come in and have access to young people.

SR: I don't want to minimize that issue at all.

TL: I hear you.

Sean, now a little gun-shy, seems to have received the message. In the next exchange, he does get Trisha to talk about some other aspects of sport culture that place athletes at risk. Coaches, athletes, and sport psychologists usually love the term "mental toughness." Sean and Trish begin to unravel what that often-valued, and often-unquestioned, quality of mental toughness actually means.

YOU HAVE TO BE MENTALLY TOUGH: CULTURAL VALUES THAT JUSTIFY ABUSE

SR: I think I'm trying to broaden the overall issue of abuse at emotional and psychological levels. It is shocking that between 10 and 15% of all athletes are being sexually abused within sport as well as all the other athletes who perhaps aren't being sexually abused but are experiencing some sort of coercion or pressure or psychological abuse.

TL: I think the other issue that we need to look at as well is some of the cultural norms, within elite sports in particular. Many are quite harmful, and I think as psychologists we have contributed to this problem as well with our concepts of "mental toughness." With coaching behaviors of bullying and harassing, well, this builds "mental toughness," and I think this is a very harmful norm. Mental toughness basically becomes defined as never complaining, tolerating high levels of pain, tolerating high levels of stress, and never showing any emotional response, and basically putting up with things. So that's extremely disempowering of the individual who is on the receiving end. It doesn't create mental toughness. It creates dissociation; it creates harm.

SR: Yeah, I thought of that when you talked about dissociation before. I thought that if someone's got a great ability to dissociate then that could lead to serious risk for injury because they'll just block it out.

TL: Absolutely, and I think psychologists are individualizing and depoliticizing systemic issues like harassing behaviors and abuse, and defining them as a problem of an individual not being mentally tough enough. In that sense the psychologists are colluding in this abuse. The whole idea of mental toughness can be positive in terms of the need for intensive training and the need to tolerate the sometimes quite severe by-products of extreme physiological training. That in itself is not bad, obviously, and so there're some parts of that concept of mental toughness that are quite useful, but the way it has been misused to allow forms of harassing and bullying behaviors on the part of some—clearly not all—coaching staff is a problem.

SR: And harassment from other athletes as well.

MEN IN SPORT REVISITED

Sean and Trish now return to the topic that was touched on, but not explored, near the end of the first interview: men in sports. Trisha's point is that the dominant model in sport and Western culture idealizes the heterosexual, able-bodied, tough, and masculine male that is embraced in the sporting worlds of both genders.

TL: I think the simplistic mental toughness concept is particularly problematic for men because it legitimizes what men have already been socialized to accept, which is that they, of course, are the mentally tough human beings. They, at least, are the dominant "species," according to what we would call the hegemonic model of masculinity, which includes being able-bodied, heterosexual, tough males. The subculture of elite sports in many ways discourages men from finding appropriate pathways for emotional expression that would allow them to be able to evaluate themselves and their responses to training, and their responses to overload, and to be able to critically assess and evaluate the other emotional stresses and strains that are going on in their lives. In other words, the hegemonic model of masculinity, which is the dominant model in some of our cultures, socializes men to suppress or subvert those emotional aspects of their lives. I think that makes a problem for men. Even for men's health, it is well known that generally men will not seek help until there's a serious and obvious issue, and we sometimes find the same thing with athletes. We have to do a lot of education for male athletes to facilitate or allow them to seek help earlier rather than later. Now this sort of environment is one that the coaches and all the other support staff need to be able to set up, one that rewards help seeking, an environment that rewards the appropriate holistic expression of emotions, and doesn't define such behaviors as being weak or lacking in mental toughness. I think we have a long way to go before we reach that kind of empowering type of environment for men in sport.

SR: Yeah, this issue was one that you brought up last time, and we didn't actually expand on it. What I wanted to come back to was this issue around silencing of men's distress in sport in general and also around the topic of overtraining.

TL: I can think of lots of examples of silencing for men around issues of identity, sexuality, and vulnerability in general. In one example, I saw where the coach encouraged a young man to talk about his vulnerabilities and fears for the upcoming event. What happened later was that the coach got really angry and called a team meeting, and all of the information that he had individually encouraged the young man to tell him, he suddenly revealed it all to the group in a betraying, sarcastic, down-putting way. The young man just felt absolutely betrayed and learned a lesson very quickly, which is: Never, ever trust another man with your vulnerability or your emotions, because it will be used against you, and you'll be punished for it.

SR: And how about the actual potential for physical harm with respect to injury when they've kept silent about whatever's going on for them?

TL: I remember one athlete in particular who had some injury-related pain, but everyone was expected to be in pain, and to broaden this up more towards our social environment, the media colludes in this "toughness" as well. If an athlete with a broken leg wins the medal, then that's a much better-selling story. That's what heroes do. The intolerance of "weakness" and the subtle, and not so subtle, messages around being strong are coupled with the feminization of any so-called

> **" I remember one athlete in particular who had some injury-related pain, but everyone was expected to be in pain... "**

weaknesses. So men get called names that are basically insults to women if they're not keeping up with the male norm of toughness. With this athlete, he said he became so badly hurt that he was out of the upcoming competition anyway. As we were talking about it, he said, "Well, what I did was I tried to ignore the pain. I just hoped that it would go away or wait until a more obvious symptom, like if a bone was coming out through my skin, then everyone would see, and then I wouldn't be accused of being a wimp or not being mentally tough." And he said, "I knew it would affect my selection if I brought it up too early because they would just see me as a whiner, and I'm not tough enough. I tried to wait it out, and then look what happened, and now I'm completely out of action for the next few months." That was a completely preventable injury, but it shows the power of the system. When you're in that environment, you have to understand the power of that environment when everything is colluding to support those norms.

SR: I imagine that sort of masculine ideal is not unique just to men in sport.

TL: I think that it's less questioned for men. It's okay for men to get together in groups and talk about sport and things like that, but to talk about relationships or emotions or give each other that kind of emotional support, that's not normed in the hegemonic masculinity model. Not to mention, very few men can actually meet those masculinity standards. Also, there are connotations for men if they are openly working on developing their emotional life or relationship life, one being the negative connotations of homosexuality with the implicit assumption that homosexuality is something negative, and one is less than a "real man." This is the harm of this dominant-model masculinity. But also, silencing around homosexuality, and around sexuality in general, happens for both men and women in sport.

SR: How do you see that as an issue fitting into stress–recovery management of the physical aspect of sport?

TL: I would see it as another stress factor where you have to be vigilant all the time that your real self doesn't show through. So you have to be disconnected from yourself or dissociate from yourself to be able to survive an environment that denies an essential part of yourself. So if you're having to disconnect and be vigilant and . . . watch all the time that you're not betraying your real self, then there's a lot of suppression from a psychological point of view going on. One might hypothesize if you're dissociated in one aspect of your life, dissociation kind of trickles through, and suppression trickles through. You're not going to be so aware of all the other things going on or the energy that it takes to keep it all suppressed. In training, extreme pushing as hard as you can and using up your distress that way, so that the true self doesn't leak through, could easily lead to stress–recovery imbalances.

THE POWER OF EARLY EXPERIENCES

SR: In the example that I gave you earlier, an athlete finally got a coach who was encouraging them to take care of themselves, to sort of look at things more holistically, but the athlete had gone through so much earlier that she still overtrained. What sorts of things have you seen from that perspective in terms of athletes starting with an abusive experience. Or the other way around, where athletes have actually come from more of a healthy background, but may have been taken down that path of overtraining because of their current experiences.

TL: I've seen it happen in both models of what you've just described. In the model where athletes have, already in their lives, experienced a lot of abuse and have internalized and normalized that as a way of surviving and then they meet a really good and empowering coach, that can be a life-changing experience for them. The potential of our sporting systems is to provide that alternative world view for the athlete, an alternative view that says, "You're not a person with no entitlements. You are an entitled person. You're a good person. You're a good child, and the abuse that happened was not your fault, and here you can be in an empowered environment." But it takes a long, long time to undo that harm and to unlearn those lessons of survival, because what the abused child will have learned is that you can never, ever, ever predict when the perpetrator is going to strike again, or who the perpetrator is going to be, so you

have to be vigilant all the time. It's never safe to let your guard down. So unlearning something that's been your experience of 10 or 15 years is not going to happen in a season. It's going to take a lot longer, and this is where I would say that we need to take seriously our responsibilities as a system to make sure that this is the kind of environment we are providing for young people who may have been harmed outside of the system.

SR: It actually makes me think of the question I asked before, which was what about the more subtle behaviors, because in some sense even coaches who have an agenda to empower their athletes can still slip up in their ways of reinforcing certain behaviors that might tap into some of those athletes' issues.

TL: I think at least some of our elite sports systems really do operate on more of a biopsychosocial framework where we have many levels of support available. We are in a better position because coaches have access to other expertise and not just their own experiences. So they get different world views of the scientific manipulation of training variables, and we hope the coach and the scientific support staff are sending out the message that training is all about precision. It's not about "the more you do the better." It isn't a guesswork game; it's actually a very precise way of manipulating your body to be able to achieve what you want to achieve. So I think that, together, coaching and training and sports sciences and all the other social supports around are in very good positions as long as they are on the same page. And then coaches are such a central person in the life of the athlete. We really need to do a lot better job of empowering coaches with the core competencies and skills that they need to be able to be effective in their roles. I think for too long we have assumed that because the coach has been an athlete him- or herself that therefore they automatically know what good coaching is, and that's not necessarily true. Sometimes what happens is that bad traditions just get perpetuated into the next generation of young athletes, and they coach how they were coached. I think a lot of our systems are now addressing these issues, even to the level of the International Olympic Committee. The IOC has just recently set up an expert panel under the Medical Commission looking at producing a position paper on abuse and harassment in sport.

SR: It sounds like coaches need a change in the way they're trained that includes skills that have not been a part of the training so far.

TL: Yeah exactly, because there's no point for us to say to coaches, "Well, don't do this yelling and screaming," because then they're going to say, "Well, what do I do then?" So we need to equip them with the competencies and skills about the ways of instructing and coaching and empowering and motivating that also work, but don't involve this dehumanizing yelling and screaming that some people do.

SR: It seems likes there's just so many avenues that we could go down.

TL: Exactly, I know you called the original title of this chapter "Sport Hurts." I don't think sport hurts. I think sport systems can hurt. Sport to me is, or can be, a very positive empowering space. It's the systems that we set up around us that are going to be harmful, and I think that's the level we need to address. [We changed the title to its current form in light of these comments.]

SR: Do you have any particular stories about athletes, good ones or bad ones, or transformational stories that would fit in a chapter like this one?

TL: I have to think about that because all I can think of are the stories where it really shows how our systems failed people in their time of need rather than supported them. Even the athlete I mentioned to you previously who came in and threw the gold medal on the table, we worked for months and months around the issues to help that athlete come to a place of valuing the medal and be[ing] able to move on and see with new eyes a perspective or an identity that was valuable rather than just beaten up.

PERFECTIONISM AND IDEALISM

In this final section, Sean and Trisha revisit several of the issues they have covered in the two interviews and touch on the problem of perfectionism. Sean, maybe not exactly an apologist for sport, is still enthusiastic about sport and its potential to influence athletes' lives in positive ways. Trisha, one of sport systems' harshest critics, is also an idealist with hopes of helping make sport a healthy and human place.

> **" ...we have to be fearless about speaking up about these things... "**

SR: Returning to what you said earlier—sport doesn't hurt, but it's perhaps the systems that damage. I still think that sport is fantastic, and I love it, and enjoy it, and hold it up there. It's looking at ways to make it more balanced, to make it so that you can pull out the positives from it.

TL: Right, and that's why I think that for those of us, then, within the system, we have to be fearless about speaking up about these things, because we're coming from within the system and coming from a place that values sport and wants the system to continue, to be healthy, and to function well for everyone. We're not coming from a position of wanting to destroy the system. I think we have to keep projecting this message and promoting welfare and empowerment in our sporting systems.

SR: I found your comments about mental toughness really quite interesting because that is something in our profession, as sport psychologists, that we get asked about all the time. It might be a request from a coach that we've got to "improve the mental toughness of the team." I don't know if it's a matter of using different terms to sway people away from this "toughness" stuff. I've had a few discussions with some other sport psychologists here in Australia, and we talked more about *resilience* because it sort of takes the image of toughness out of the equation.

TL: I think we need to critically evaluate concepts and attempt some definitions that reflect well what we mean by health-promoting mental toughness or resilience as opposed to pathological resilience or suppressive mental toughness. I agree we need to clarify, to unpack the definition of what we mean and identify well what are the behaviors that would indicate that someone is being mentally tough.

SR: How would you describe it in the negative sense, and then how would you describe it in the positive sense?

TL: I think the positive sense is probably more in line with what you're saying about resilience. Someone who can rise . . . See! Even beginning to talk about it, we get trapped in our words, and in our language, and in our definitions.

SR: I guess maybe we could look at behaviors that reflect the positive side.

TL: I think we have a lot of examples of athletes who do embody the positive side and who are willing to do what it takes to achieve excellence in their field. The problem with the mental toughness concept is that it kind of implies a head-down-go-as-hard-as-you-can-and-close-off-everything-to-reach-the-desired-end [type of approach]. It's a problematic and simplistic concept. We need to throw it out and start working from a point of view of what kinds of behaviors are helpful for someone who wants to achieve excellence.

SR: What comes to mind right now for me is my experience working in the ballet. In ballet, they talk about the need to be *absolutely perfect* and that creates all kinds of pressures similar to the demands of mental toughness. I talk to dancers about using their perfectionism in a positive way, and in that sense, like with mental toughness, be tough about your nutrition, be tough about your balance, and be tough about your recovery. I sort of say, "If you want to be perfect, strive for perfect balance than rather than for some sort of illusory perfection."

TL: Those are excellent concepts. I think that's a really good way of putting it: Be tough about doing the right things and not compromising on the right things.

SR: Like you were saying before about silencing, it takes a lot of courage and a sense of toughness to speak up and not to just deny what's happening.

TL: Exactly, and to acknowledge the humanness of your life, and despite your vulnerabilities and frailties, despite all of that, you're still doing these right things. That's true excellence. Pretending I have no vulnerabilities, and therefore I'm a super person—I mean, this is not realistic.

SR: Well it'll be a nice day when we can sort all of this stuff out.

TL: I know, I know, and the world will be perfect.

SR: And there it is again: perfection.

TL: Exactly! Oh my goodness.

SR: The discussion about mental toughness certainly was something that we really didn't get into last time, and I've seen, now that I've developed a bit more as a practitioner, that perfectionism, and whatever the hell *mental toughness* is, are major issues in sport and major issues for sport psychologists.

TL: I agree, and maybe those topics should be examined in some further discussions.

SR: Sounds like a plan. I guess we'll end here, and thanks so much for your time and thought-provoking comments and stories.

REFLECTIONS ON TRISHA

The content of this chapter is, to say the least, rather in-your-face. We knew Trisha Leahy would be controversial, confrontational, insightful, maddening, and sobering. Trisha is telling us that in mainstream sport psychology, with its focus on performance, the intersection of research and practice with the issues of overtraining is limited and limiting. We know that overtraining is highly idiosyncratic, and where we really want to *start* looking is at the level of the individual and how that person functions in the dysfunctional and potentially damaging world of sport. We also need to understand the athlete in the wider sociopolitical context. When we stay focused on individuals, it's like taking a Diego Rivera mural representing the economic, social, political, religious, and familial struggles of Mexican peasants and focusing on one figure who appears sad, plucking that figure out of the mural, and saying, "Here's overtraining." The focus is too narrow and misses the larger, more complex story. Past researchers have been looking in the wrong places. Or maybe they were looking at, and trying to help, a sick tree but didn't notice that the whole forest is infected.

Trisha's main point is that we need to look at both the individual *and* the sociocultural milieu, along with all the media and financial pressures that come to bear on athletes. Her points about abuse on the playing fields that would not be tolerated in the classroom should strike a chord with parents, who usually want the best for their children but, like psychologists, can fall into the bystander position and contribute passively to the continuation of abuse.

Our blinkered approach to sport psychology service, with its dominant focus on the individual, can be shown in other areas of research in our field.

For example, in the areas of leadership and task- and ego-orientation research, there is a question of focus on individuals having certain leadership styles or athletes having task or ego orientations. But where did these "traits" come from? If we take Trisha's approach, we begin to see that these traits have roots in the larger social-political-cultural context, and trying to help an athlete develop more task orientation through something like cognitive restructuring may be doomed to failure if the sociopolitical climate reinforces winning at all costs. Trisha's comments go far beyond abusive relationships and can apply to a whole host of other issues we address in research and practice.

A CLOSING STORY FOR PART II

To end this section of the book, we would like to tell another story from a long time ago. There are many amusing parables in the 13th century Sufi tales of the Persian (sometimes Afghani, sometimes Turkish) Mullah Nasruddin. He is a comical character, and his brief tales are told across the Islamic world as jokes. Nasruddin parables, however, usually have much deeper meanings and can be used to help us understand the human condition. One story, in particular, illustrates the problem of how we look at overtraining.

One day Nasruddin lost his ring in his house's basement, where it was very dark. There being no chance of his finding it in that darkness, he went out on the street in the daylight and started looking for it there. A friend passing by stopped and inquired, "What are you looking for, Mullah Nasruddin? Have you lost something?" Nasruddin replied, "Yes, I've lost my ring down in the basement." Somewhat confused, his friend asked, "But Nasruddin, why don't you look for it down in the basement where you lost it?" As if it were self-evident, Nasruddin said, "Because there is more light out here. How do you expect me to find anything in that darkness!?" We look for answers where there is more "light," where exploration is easier. We don't go into the dark places.

Trisha's point is that, like Nasruddin, when we look at overtraining, we are looking out on the street in the light. Out there we may find many interesting things (e.g., perfectionism, serum cortisol elevation), but we won't find the key to overtraining. It lies in an area of darkness that we don't explore. We can't see that it is the whole pathogenic, sociocultural sport system that is the subterranean foundation (basement) of overtraining and many other problems. Sport systems' underbellies are in the darkness because they are occluded by our unquestioned bias that sport is good, that what we do with athletes is good for them and builds character, and that sport brings out the best in people. Sometimes these prejudices hold under closer scrutiny; many times they do not.

As we said of David Martin, the field of sport sciences needs radical thinkers and practitioners such as Trisha Leahy. We need to be challenged, shaken from our prejudices and dogmatic slumbers. David and Trisha are seasoned researchers and practitioners who are at stages in their careers where they can give something back to their respective disciplines. Their gifts are not simple; they are complex, confusing, and confronting. The most valuable gifts often are. We would like, again, to thank them for their gifts. The question now is: How will we receive them? Will we say, "That's nice," and put them back into a cupboard? Or will we try to understand them, examine them closely, and attempt to use them in our practice?

Policy Documents From the Australian Sports Commission

Policies on a variety of issues covered in this chapter can be found at the following Web sites:

- Behavior and sport:
  www.ausport.gov.au/policies/behaviour.asp
- Child protection:
  www.ausport.gov.au/ethics/childprotect.asp
- Harassment-free sport:
  www.ausport.gov.au/ethics/hfs.asp
- Junior sport:
  www.ausport.gov.au/junior/jsf/index.asp
- Women and sport:
  www.ausport.gov.au/policies/women.asp

WHAT CAN WE LEARN FROM ATHLETES?

The following three chapters are told from the viewpoint of Sean, who conducted the interviews. The original group of 13 athletes I (Sean) interviewed came from various sports, representing different emphases in training and competition requirements (i.e., aerobic versus anaerobic, and primarily skill- versus effort-focused). When selecting participants, my main criterion was that the athletes had a history of chronic overtraining, as identified by the experts in chapter 4.

As I began analyzing the interviews, I started to feel that the athletes' tales had a qualitatively different feel from the stories told by the experts, and that an inductive content analysis (identifying major and minor themes) of the athletes' histories would probably lose the most valuable parts of the athletes' interviews—the detailed experiential elements. After discussion with my coauthors about the differences between expert and athlete reports, we decided to present the stories from the athletes in a different format. We agreed that I would present the athletes' interviews as aggregate case studies, combining the interviews to create a smaller number of fictional narratives based on the actual lived experiences of the 13 athletes. This structure allowed me

to minimize repetition and overlap in telling the athletes' tales. It also allowed me to give rich, detailed accounts of athlete experiences with OT, while protecting the identities and confidentiality of the participants. I discussed the athlete profiles with Tony and Mark, and together we came to agreement on three distinct stories of fictional athletes drawn from the interviews:

- A professional footballer driven to abuse his body by a relentless machine of economic forces, internalized clichéd maxims, tough sport cultures, and traditional abusive practices
- A triathlete driven to overextend his body in pursuit of Olympic gold
- A young gymnast-turned-cyclist driven to distort and damage her body by abusive coaches and a pushy, overinvolved mother

PROTECTING CONFIDENTIALITY

Within each of the three aggregate stories, I was careful to alter any details in the verbatim quotes that might identify the real-life athletes whose experiences were the subjects of the tales. For example, to protect the identities of the athletes, I generally chose the sport of participation for the fictional characters from sports not represented by the actual athletes whose quotes contributed substantially to that story. The one exception was the footballer, whose identity and confidentiality I felt were already protected because of the large number of athletes participating in that sport in Australia, as well as the lack of other identifying details in his quotes. The fictional element is particularly evident in the story of the triathlete. For example, triathlon was not included in the Olympics until 2000, yet I tell the story of a triathlete competing at both the 1992 and 1996 Games. I also altered potentially identifying details such as competition dates, numbers of world titles, locations of big competitions, and times of the season.

In the next three chapters, then, we present the three fictional narratives, which are the aggregated case studies based on the 13 athletes interviewed, beginning with the tale of Steve, an Australian Rules professional footballer. In the final chapter of part III (chapter 10), I tell my own tale of overtraining, part of which was briefly presented in the book's introduction. Then, at the end of chapter 10, we offer observations and reflections on my and the three composite athletes' tales.

CHAPTER 7

THE PATHOGENIC WORLD OF PROFESSIONAL SPORT

Steve's Tale

When I (Sean) sat down with Steve, he was immediately friendly and eager to share. He was here to tell me his story of how he felt compelled to abuse his body, trying to live up to the expectations of the footy-mad public, fighting to withstand the burgeoning economic pressures in professional sport, and struggling to overcome injuries in one of the roughest team sports in the world.

INTRODUCING STEVE

Steve had retired 2 years before our interview from a successful professional career in the Australian Football League (AFL). He had played 200 games in 11 years of senior football and had been on a premiership-winning team—the holy grail of the AFL, equivalent to the Super Bowl in American football. Yet there was a subtle sadness about him that would pervade his story of overtraining (OT) and injury. Steve seemed to have some regrets about his career in the AFL, and he wished things could have been better. He thought that maybe he could have left the sport with much less damage to his body, or ended his career more gracefully, if he had not been driven constantly to play with injury.

Steve grew up with a family legacy of successful athletes—grandfathers, grandmothers, mother, and father—who were national-level athletes in their sports. I had the sense that this legacy had a strong influence on Steve's football career (more on this a bit later). Steve seemed to have followed his family's traditions and responded to its pressures. He took to the sport of football at age 6 and never looked back.

EARLY AMBIVALENCE ABOUT SPORT

What struck me early were Steve's comments that he did not have an overwhelming ambition to play professional football.

Steve (S): I've never had an ambition as a kid actually to play footy. My parents . . . I thought they probably pushed a little bit to play, and they, obviously, like most parents, drive you here and there. But it was more like a progression thing than an ambition to play top-line sports, like you just did it because you're probably good at it.

It seemed odd to me that Steve reported both a lack of ambition and a strong desire, when many athletes might recall having dreamed of reaching the pinnacle in their chosen sport from a very early age. Steve did not go into detail here about the role of his parents in his motivation to play footy, but I took note, filing this comment away, waiting to see if he might return to the issue of his parents' influence later in the interview. I did find it intriguing that Steve seemed to downplay his ambition because it went against many of the stories I had been hearing. It was as if he just fell into the sport because he happened to be good at it.

S: So naturally you do things that you're good at because it probably makes you feel good or whatever. So I never had ambitions to play league footy. It was just, "Okay, I'm at that level, so, oh, that's what I do." Then you get drafted and play at top level in South Australia, and you think, "Oh well, might as well go to the VFL (Victorian Football league, one step below AFL), try that," which came to the AFL. You think, "Oh well, that's what you do," and you just play. So, it was not really ambition.

I wondered how these comments about lack of ambition and potential parental pressures fit in with the overtraining issues that arose in Steve's career. Maybe Steve was a victim of others' ambition, or maybe he was a victim of his own ambition to win the love of others by being a great athlete. It made me think about that bit of sadness I had detected at the beginning of the interview. Why was this successful guy, with an enviable history in professional sport, not really all that happy as he reflected back on his career? Why did he look back and question his own motivation or ambition? I wondered if distancing himself from ambition to play professional football helped him cope with the disappointment he faced during his later years in the game, years spent battling injury and overtraining, and enduring public scrutiny. I also began thinking that a professional athlete's career is not as enviable as it might first seem to the outsider. I was keen to hear more of Steve's story about his journey through the pathogenic world of professional football.

LOVE OF THE GAME AND REGRET

Despite his apparent lack of dreams and high ambitions, Steve did love the game. He described some of the things he missed by not playing anymore, such as the rush of playing in the big stadiums, the thrill of hearing the screaming fans.

S: The adrenalin. . . . You're used to running out on the field . . . and it's an emotional roller coaster. You're up; you're down! Then you go from running out in front of 60,000 or 100,000 people to no longer playing anymore. It's like, "What do I do now to give me a kick?"

He also voiced fears about how life would be without football.

S: One day your career could be over . . . and if your whole life's just been focused around being an athlete, all of a sudden it's gone. So where do you fit in the world? You've no longer got the adulation or the people talking to you and wanting to do things for you anymore. Two years after the game, it's like, "Steve who?"

Steve's comments about the fear of life without football made me think of what some experts in chapter 4 had mentioned about athletes being at risk for overtraining near the ends of their careers (e.g., wanting to go out on a high note, feeling anxious about losing all the adulation, training hard to stay in the game, training hard in response to potential future loss of identity). It seems that professional athletes sometimes get caught up in pushing themselves too hard as they fight to prolong their careers, clinging to transient fulfillment from the adulation of fans and from the adrenalin rush of being a performer on the big stage of sport. Despite claiming little initial ambition to pursue football, it sounded like Steve had developed an identity around being a professional performer, being a person loved by thousands of fans. In becoming injured, however, Steve descended from being *somebody* into a state of identity panic, as the loved and admired identity slipped away. I have heard other athletes, coming to the ends of their careers, describe their last few big events as exceedingly important to them. I have also heard numerous times, in the media, of athletes announcing that they want to retire on a winning note, preserved in the memories of the public as champions, as winners.

S: I knew it was the end of my career. Going out on a good note—that was the last thing I wanted to do. . . . I just had things I wanted to achieve, and then I wanted to move on to my next life, and I'd always said that. I didn't want to just be in it for the sake of being in it.

Unfortunately, as noted in chapter 4, athletes facing such transitions often resort to excessive training. To train harder and longer becomes the perceived antidote to their fears of being forgotten; of fading away; of being remembered for weakness, injury, and poor performance; or of being nothing. The problem is that their aging bodies often cannot tolerate the strains of the overload training; they require more recovery than they did at younger ages. Ultimately, overtraining does not work well as a coping mechanism for career termination anxieties; it might even be compared to using alcohol to help cope with emotional pain. At first it works, but then it becomes part of the problem, in the paradox of self-destructive coping. Overtraining behaviors cause the athletes' fears to be realized, as they descend further into states of performance decrement, prolonged fatigue, injury, and illness.

> **" When you're on a contract and they are paying you good money, they expect you to play. They don't care that you are injured. They want you out there playing. "**

For Steve, however, the pressures to train through injury, to push harder, and to neglect recovery were not just isolated to this transition period at the end of his career. Steve expressed the feeling that he was caught in a manipulative machine that did not allow for individual differences, that dismissed any possibility of human weakness and the needs of the human body for recovery, that praised gladiator-like performances where players limped onto the field and played despite severely damaged bodies, and that demanded entertainment to appease the appetites of a sport-crazed public. He was trapped in the relentless machine that is professional sport.

I could not help but think, "Hey, at least you got paid to do something that you loved to do!" I have known many amateur athletes, including myself, who have chased sport glory for many years and had nothing more to show for it than wounds, scars, and chronic injuries. I realized, however, that I was having a rather narcissistic response to Steve's story, probably stemming from the feeling that I did not reap the rewards that Steve did from sport.

PLAYING FOR PAY

Nonetheless, more often than not, the issue of getting paid for performance seems to be a major risk factor for overtraining. Steve made it clear that contract issues, and the financial pressure involved with being a key player in a multimillion dollar industry, often prompted poor decisions from coaching staff and others about athletes' health and readiness to play. Steve told a story of how being out of contract (his contract not being renewed) put him at risk for overtraining and injury.

S: Before I went to [the new club], I was out of contract, couldn't negotiate. I didn't have a contract, so had to organize another club to go to, and I missed the first 4 weeks of the preseason. . . . Did 4 weeks training, then had Christmas, and then they started playing games after Christmas. My management team didn't want me to play because I wasn't contracted anywhere. There was always a chance of injury, and the club didn't want me to play in practice games. So, I wasn't doing all the work, because the club I was training with, their preparation was basically playing games. So, I didn't actually do a lot of training, and then I got picked up by [the new club] and, at that point of time, I had missed out on probably 2 or 3 months of really hard training. So, you are behind the eight ball to start with. You haven't got match fitness. You haven't got the capacity to run the whole game, and then I got injured.

Experts in chapter 4 pointed out that training too hard after a break puts athletes at greater risk for overtraining and injury. It appears that Steve did not have much choice, because of contract negotiations and the need to get himself physically prepared for the intensity of match play after returning from the off-season break. He had not trained enough for match play and was underprepared for the demands. The demands exceeded his resources, and by training and playing over his limits he incurred an injury. There seems to be no doubt that money is a big motivator for most people involved in pro sports. Steve was aware of the mighty dollar's influence: "There is always that financial incentive to come back early and actually play, and then you do."

Steve also commented that the pressures of a professional contract, often associated with players coming back too early from injury or not communicating about injury or fatigue, can be different for players in different situations.

S: They've got some people on fixed contracts, and then they've got kids who get paid per game. "Play him. We've already got to pay him this amount of money." Rather than,

"If we play this kid we've got to pay more, because he's on a per-game basis, so we lose twice." When you're on a contract and they are paying you good money, they expect you to play. They don't care that you are injured. They want you out there playing. If you're on a fixed-fee contract, which most of the top players are, they get paid a fixed fee for the year. If you're injured; you still get paid. Therefore, the pressure is on you from the president of the club on down.

On the one hand, athletes getting paid on a per-game basis could be motivated to keep silent about any niggles, injuries, illnesses, or fatigue, because they want to earn the bonus for playing that game. On the other hand, players on fixed contracts get the pressure to play from management because the team is trying to cut costs. In effect, there seem to be different pressures on different players, but every player ends up feeling pushed, possibly far enough to engage in unhealthy training practices. Having worked for an AFL team, I have seen players express contract concerns. During times of contract uncertainty, some players seem tempted to push through injuries, or to choose not to disclose fatigue or soreness to the coach, especially junior players hoping to get offered contracts to be full-time senior players. For Steve, the money pressures were substantial. Once again, I caught myself thinking, "Being a professional athlete is not as fantastic as it seems to the outsider." I could see how society has created an image of the supreme physical being, the richly rewarded professional athlete. It is an image to which the athlete becomes attached, an image that is constantly reinforced by the media, by the public, by the coach, and by the athlete's own needs to form an identity. Steve was definitely caught in the pressure cooker of professional sport, and it seems that he saw no escape. He commented, regarding contract pressures, that one cannot escape even the scrutiny of one's own teammates. There's also player expectation: "Oh, he's getting paid a lot of money, but he's not working hard." Or, "He's getting more money, and he's useless."

INJURY MISMANAGEMENT

At this point in the interview, I was beginning to get a picture of some of the significant overtraining risk factors inherent in professional sports such as the AFL. It seemed that one of the major themes coming out of the interview with Steve was that injury is inevitable in the AFL, but poor injury management, often prompted by financial pressures and constraints, paved the way for all sorts of overtraining issues. Injury management (or rather mismanagement) was an area of Steve's sport where there was considerable reinforcement to neglect recovery, to keep silent about pain and fatigue, and to play while injured. The financial pressure was just one of the factors prompting footballers, as well as their coaches and other training staff, to make poor decisions about training and recovery.

I thought back to when I had started this project with a one-dimensional perspective on overtraining. I had understood overtraining as something consisting of a serious fatigue syndrome, called overtraining syndrome, which probably did not affect most athletes, and which was likely to be restricted to endurance sports with heavy training loads. Some authors in the overtraining literature had commented that injury could be an outcome of the process of overtraining (e.g., Kibler & Chandler, 1998; Kuipers, 1996), but several others had not included injury in research on or definitions of overtraining (e.g., Raglin, 1993). What I saw with Steve's story, and, as it turned out, many other athletes' stories, was that injury outcomes are closely related to overtraining behaviors. Injury can be seen to develop from excessive training, from neglecting recovery, and from stress overloads of any type, including social and emotional stress; injury might be the most common outcome of overtraining behaviors across all sports.

Steve talked a lot about injuries; they seemed to rule his life at times. The sadness that was part of Steve's memory of professional football appeared intimately connected with his injuries. From the way Steve described his experiences, he was almost constantly living in a damaged state, and there never seemed to be adequate time to rehabilitate fully from injuries. Steve said that as an injured player he had the perception that he was always behind the rest of the group, that he was always trying to push himself to do more to catch up, to make up for what he had missed.

S: One year I came off having two knee injuries that I had had from the previous season, so I had operations on both knees, and therefore I didn't really start training until January, because of the rehab and all that sort of stuff. So I didn't start running until January, and I had missed out on all the 2 months prior to Christmas. Then you've got another month of solid running before you even rejoin the group. So, you're down on touch and kicking and skill work, and you haven't got that 3 months of extra base, because you couldn't run, and couldn't even cycle. You never catch up because it's a fine balance. The more you do, the higher the chance of injury. So, it's a very fine line between doing too much and not enough. When you come through that, you try and catch up. You try to play the games, but you are not prepared to play the trial games because you haven't done all the work, but you play because you've got to get some fitness. Then, coming off being behind, in the preseason, I strained a quadriceps. Therefore, I'm out for another couple of weeks with that, and I'm getting further behind in my fitness because I'm not actually out there training. Then, coupled with that, on the Wednesday before the first game I got glandular fever (mononucleosis). So that topped the whole thing off, and it was out the door.

COACH AND MEDICAL STAFF PRESSURES

From this account, I could see how feeling behind seemed to set off a chain of unfortunate events and maladaptive responses, ending with Steve getting dropped from the team. The perception of always being behind, of always trying to catch up, can be a powerful motivator for an athlete like Steve to push too hard. Yet doing so—pushing to play despite pain, injury, and illness—is not just a response to feeling behind. Steve talked about how coaches and other staff pushed athletes to play with injuries.

> **S:** It's a lot of pressure, but then the coaching staff also forgets. They know you're injured when they're in the calm of a normal week, but under the intensity of match conditions, they see you not being able to do something, and, all of a sudden, they say, "What the hell? Go out and tell him. Give him a rev." You know you can't do any better because you physically can't do it, but they forget because they know what you can achieve, your high level of performance, and, if you're not performing to that, there are questions asked, and

then they get different opinions of you, and all of a sudden one injury, especially for a young kid, can make or break their career.

Listening to Steve talk about the coaches' demands, and lack of acknowledgement for his struggles with injury, made me feel irritated, and that irritation, in part, probably stemmed from my identification with Steve, from interpreting Steve's experiences through my lens. What are these coaches thinking? I know from my own experience, and from other athletes I have seen, that most athletes really would like to continue playing, no matter what. It seems unjust for a coach to question Steve, to judge his ability while he is injured, just because the coach is too caught up in his goals of winning the match (and keeping his job). Steve voiced his frustration at his coach's blindness to the situation.

> **S:** Having the coach demand that you play even though you were injured. . . "God, coach! There is nothing more that I would want but to be injury free and playing 100%."

Steve protested that the coach did not see the situation for what it was—that he was a player limited by injury. Unfortunately for Steve, he was caught in a culture of hypermasculinity, and, in that pathogenic system, being injured was, like being castrated, an emasculating experience. Steve acquiesced to the demands of this "be a man" culture. Those pressures led Steve to make choices that damaged him. "Choice" is an odd word here. Given his choices, it seems like he had only one, and a poor one at that. The exploitative and damaging sport system, and the pressures involved, are exactly what Trisha Leahy spoke about in chapter 6. Steve was disempowered and trapped within an abusive world, yet personally driven to stay and be a part of that world. As a result of the coach's actions, Steve claimed that he was left with a dilemma: Play to please the coach and risk more serious injury and poor performance, or sit out and risk the scorn and judgment of the coach and others. There is really no dilemma, no choice—just a path that leads to physical damage. The other option won't work.

> **S:** You could play up in the senior ranks for three games, and then go out and play with an injury because you are young.

> *" **Having the coach demand that you play even though you were injured. . . "***

You think, "Oh shit! How do I manage my body better?" And you've got, say, a groin injury, and you go out and play, and then, all of a sudden, you play two bad games. You get dropped. Your groin's no better. End of the year, and you're still injured. "Nope, he's no good. Get rid of him." Because all they remember is the last game you played. The last thing in their minds is what they remember you by. And that's just the way it is, and so there's so many pressures.

Steve suggested that, for young athletes, the answer to the dilemma usually is to please the coach by getting out there and playing. I had the impression, however, that Steve was not limiting this dilemma to young players. He seemed to be saying that he had faced this dilemma throughout his career, even as a more mature player. I also thought back to Steve's earlier comments about parental pressure and wondered if a subtle unconscious link was being made here between needing to please the coach and needing to please a parent. This thought would continue to hold as Steve's story unfolded.

In addition to a demanding coach, it seemed that plenty of staff members, including the medical staff, had colluded in encouraging Steve's overtraining behaviors, especially through reinforcement for playing with pain and injury.

S: I had a back spasm on the day of the game. That morning I rang up the doctor to say I had back spasms, went and saw the doctors. Rang the coach, and coach went ballistic at me for being injured, "Oh, it stuffed up all my plans!" "Crikey, Charlie! I can't play." I couldn't walk. How do you expect to play? They just lose control. Even the doctors were saying, "Oh, we'll jab this, jab that, to play," and I'm thinking, "I don't really want to. If I'm injured, my body's telling me I'm not right to play, and you want me to play?" That's a high-risk game. A lot of times you're going to fall over. You might get away with it once, but you keep doing it. It's like a game of Russian roulette; eventually you are going to cop one with a bullet. It's a vicious cycle, and you're really only a commodity, that's all.

> **" It's like a game of Russian roulette; eventually you are going to cop one with a bullet. It's a vicious cycle, and you're really only a commodity, that's all. "**

Steve presented an image filled with disillusionment, expectation, and sadness: "a commodity, that's all." I sat there looking at Steve, imagining what it might feel like to be seen as nothing more than a commodity, a thing, not much more than a bottom line on a balance sheet. In the face of serious, limiting injuries, football players are being asked to get on the battlefield, to face the bone-crushing tackles that are part of the game. Sadly, Steve's support system consisted of a doctor with a syringe full of corticosteroids, the drug of choice to distract the athletes from their pain. Furthermore, it sounded like the frustrating aspect for Steve was that he could see, quite clearly, that he was doing the wrong thing for his body. He was trying to object to playing, but his objections tended to get drowned out by the clamor to play. Here was Steve with overt, direct pressures to play with injury. Did he have a choice? Did he have a voice that could be heard? It did not seem so. At times, Steve suggested that his coaches might have shown some insight into pushing too hard and not allowing enough recovery, but too often this insight occurred only after the damage was done.

S: Although there is a lot of overtraining, the thing is that the majority of the times the player knows his body better than anyone else, and it's scary, and it is like you've been there. You're feeling sore for some reason, so you tell the medical staff, but, "Oh no, you go out and train." You say, "I've got a sore hamstring." "Oh no, you'll be right. Go out and train." The medical staff is in a Catch-22. It's a fine line, a balancing act between sending a player out to play and putting him on the injured list, because the coach wants the players out there. It might not be in the best interest of the players, but the coach needs the players on the field. Because you're an elite athlete, "Okay, you've got to play with some pain." You go out there, or you do extra, and the thing is, "I know it's not right. . . . I'm a high risk here!" And when you get out there the coaches forget about your injury, and they say, "Run harder," and then you end up doing a hamstring, and it's like, "Oh, in hindsight, oh, I shouldn't have trained you."

After these comments, and the one about doctors using cortisone injections to dull the pain of an injury, I found myself wanting to hear more about the medical techniques being used to augment the coercive pressures to play while injured. Steve readily gave me several examples where injections were used to get him back on the ground. I am not a medical doctor, but it is my understanding that cortisone injections might help with reducing inflammation of an injury, which might help with healing during a course of rest and rehabilitation. Steve's stories, however, seemed to show that these injections were being used to mask the pain of an acute injury so that an athlete could continue to play, but at the risk of getting much more seriously injured.

S: I remember . . . When you play, the coaches don't really care. They just want you out on the ground. I have played with stress factures in my feet, having injections where they've stuck the needle in, didn't work, and you just come up, and coaches say, "No, you've got to play." One year I came off, and I'd been in a collision, and I'd cut my forehead open, and someone slid into my ankle, and I had to come in to get my head stitched up. You go to the doctor and you say, "Oh, I've got a sore ankle," and he said, "Oh, have a look. See if you can run up and down there." And I said, "I'm sore." So, he put four injections in it to try and kill the pain, and I go back out, and I went for a run, and I said, "No, it's still sore. I don't think I can go back on." And as soon as you tell the coach you can't go back on, at a break, the coach just walks straight past you. It ended up I had a broken ankle and was out for 7 weeks.

Using such exploitative medical practices sends a message to athletes, coaches, and others about how drugs could be expected to help with injuries, and, more important, that playing with injuries under medication is *what you do*. Such practices get perpetuated in sport when they are not questioned, and athletes might get involved in self-medicating behaviors so that they can continue to play or compete. Steve's tale continually reminded me of what Trisha Leahy had said, and I was seeing the

> **" ...as soon as you tell the coach you can't go back on, at a break, the coach just walks straight past you. It ended up I had a broken ankle... "**

product of the dysfunctional and abusive professional sport system sitting in front of me.

S: Yeah, I got the injection, and the next day, after our final preparation camp, I started trying to compete on Panadeine Fortes [paracetamol with codeine], just chewing them down like they were nothing. I couldn't do it. I couldn't jump. I'd try to take off, and it felt like a knife was going into me. So, again, had to make the decision and spoke to team management and said, "I'm out. Sorry."

I was affected by the image of Steve "chewing down" pain killers in an attempt to keep competing because it made me think of my own desperate attempts to overcome injuries with any methods available. It seemed like a futile attempt to fix a problem that could not be fixed with more masking agents. Is self-medication, masking pain and injury in the name of sport glory, a type of behavior that one would applaud? I do not think so, and Steve seemed to agree, pointing out that he was aware of what was going on around him, and that many things were not right with respect to the pressures to train and compete with injury. I could see where some of the pressures were originating, such as financial strain, but I guessed that more than that lay behind Steve's overtraining. What other factors were driving Steve to push himself too hard sometimes, to neglect recovery, and to succumb to pressures to perform when the likelihood of further damage was high?

GOING THE EXTRA MILE

I was getting a better understanding of the culture of Australian professional football and how, perhaps not unlike other sports, it accepted and encouraged overtraining behaviors. Steve did mention that, early in his career, he was doing extra training to try to get ahead.

S: Obviously, trying to get to a certain level, you've got to do the extra running and weights and build yourself up. In the past, especially in the start of my career, I used to go out and do extra running, do speed work. You find sprint coaches, and you just did it, and get with a coach or a skills coach and do extra kicking skills and that sort of stuff.

It seemed that Steve had some internal drive to do extra training, despite his many comments about the pressures coming from around him in the sport, and despite downplaying his ambition to play professional football.

S: I've overtrained. I'm one who likes to train myself and not do the group training because I think I know what's best for my body at times, because group training normally becomes so robotic, and you do the same thing all the time, so it gets boring. They don't put enough variety into training, but yeah, I have overtrained. Well, I mean, I always do that little extra bit, but I never told them that I'd do it. I don't feel the need to go and brag. What I do in my free time, they don't need to know.

The extra training early in his career has features that reflect some of the risk factors mentioned in chapter 4. Why did he go the extra mile? Was it because of perfectionist tendencies? Did it stem from internalizing the more-is-better ethos of sport? Was he trying to please significant others with his accomplishments? Did he embrace the equation *good performance = good person*? These sorts of questions were popping into my head as Steve told his story of doing "that little extra." Looking back across his career, Steve also acknowledged that this extra training was related to several serious injuries.

S: You get to the point where my last two injuries were degenerative injuries. I had a hernia, and I had to have a tendon cut in my groin so it released some stress, got holes drilled in my pubis bone because it was degenerating. Had an operation on my knee. It was a degenerative knee complaint. That's just obviously from running on roads, extra training, and doing the whole thing the whole year. You do your normal training, and then you think, "I want to do a bit more speed work," or do a bit more fitness, and you do that on top of it, and you just load the body up with more work. You don't get enough chance to rest.

> ❝ **They're taking kids at 10 years old and training them like they're professional athletes. At that stage they've got to have fun.** ❞

LESSONS LEARNED EARLY

Reflecting on Steve's internal drive to push harder and do extra training, I thought back to the influence of his parents, hinted at early on in the interview. As the interview progressed, Steve came back to his parents on his own, though he began not by talking directly about his own experiences as a young footy player, but by addressing junior footy in general.

S: So, the scary thing is that it goes all the way down to junior footy. They say, "Oh, the AFL do that, I reckon it's good for my team." It is just scary. They're taking kids at 10 years old and training them like they're professional athletes. At that stage they've got to have fun. You don't need to train them, just give them a ball and have a kick. They don't need to do 5-kilometer runs, 400-meter sprints, doing the whole fitness program. It's just ridiculous. Let them get out there, have a kick, and just enjoy themselves. I don't know why they just don't let them go out there and play. They're trying to make young people into professionals, and they don't need to.

It seemed to me that Steve was referring to his own experiences as a 10-year-old boy, how he might have felt pushed as a junior to be a professional athlete. Steve's next few comments went directly to the issue of his father's overinvolvement in his sport. Immediately, I could see where some of Steve's internal drives were coming from; perhaps it was how he as an adult footballer could have fit easily into the dynamic of wanting to please the coach, a new father figure.

S: They've got expectations, and parents are the worst. Parents! And I was probably under pressure. Your parents expect you to play and expect you to do well, and you go out there, and you play, and you know it's like, if you don't play well you're going to hear about it.

It sounded like Steve had really heard about it from his parents when he did not play well. I was fascinated, however, by Steve's guarding of obvious anger toward his parents for the pressure that he felt. He said that parents are the "worst," but then commented that he was "probably" under pressure. He also used a distancing presentation by putting his comments into the second person ("you"). The fascinating part for me was how Steve's guarding of anger is something I have seen quite frequently in my therapeutic interactions with people, as well as in my own life. I think as children we have trouble expressing anger toward our parents, because we fear that our anger will deprive us of their love and acceptance. Steve's next comments erased most of the doubts in my mind about whether he had received coercive pressure to play injured, even as a child, and whether his father played a significant part in influencing his overtraining behaviors as a senior footy player.

S: When I was a kid, I always played 2 years above my level, and I played with my brother, and my dad coached me, and I broke my fingers, and I remember it clearly. I must have been about 7, and I am playing in the Under 9s so everyone else, every kid is almost 2 years older than me, and I'm the smallest one. Broke my fingers, and I remember my dad; he was the coach. He said, "Well, if you're going to whine, get off!" I went to the hospital, and I had a broken finger.

This image was enough to make one cry: a 7-year-old, the smallest one on the field, with broken fingers, being told off by his father, the coach, in front of all the other players, for being upset about real pain. What alternative is left for the kid but to try to suck it up the next time he is out there feeling the pain? From Steve's next few statements, it sounded like humiliation was a tactic that was not used sparingly in the world of junior footy, and was not limited to injury concerns.

S: Coaches are like that. Coaches have their own egos, and I'm not saying that's my dad, but it's the whole thing, the same principle. Coaches get in that mentality: "We've got to win. We've got to win!" And the parents are the same, and the worst people are the parents. They just slag off all the time and abuse their kids for making simple mistakes. We all make mistakes, and the expectations are incredible on the kids. You see why they don't want to play sports, because the parents put just so much expectation on them. OK, it's all right to push them, well not push them, but encourage them to play something, and playing in a team

sport is probably good for kids, but parents go overboard, I think, a lot of the time. Yeah, it's an interesting one.

Steve says he was not describing his dad, but I felt that he was telling a story about his dad. The contradictions were palpable: He was defending his dad, but then saying that parents are the "worst," for the second time. It seems that he meant to say that *his* parents were the worst. Perhaps Steve really enjoyed the game of football but had the fun taken out of it by a father who always emphasized winning, verbally abused him for making mistakes, pushed him to play with injury, even at a very young age, and expressed overwhelming expectations for Steve's performances. Steve's first comments about not having had personal ambition to play professional football seemed to make more sense now. The sadness I had sensed also seemed to have sources beyond football. Although Steve had said that he loved the game, it appeared that the ambition belonged to his father. That dynamic between a father and a son can easily be transferred to a coach–athlete relationship. I imagine that, in the often abusive world of professional football, with its pathogenic foundations, it was all too easy for Steve to get sucked into that unhealthy father–son dynamic with a coach.

Many of the experts I interviewed mentioned family dynamics and pressures from significant others as risk factors, and we have discussed the problematic aspects of having parents serve in dual roles when they coach their children. With Steve, we have an illustrative (and saddening) example of how a psychologically damaging parent–athlete or coach–athlete relationship can get played out again in adulthood. The experts also noted that parents may try to live through their children and that children's potential financial rewards may influence how parents behave. Steve did not specifically mention these issues, but there is a possibility that these forces were also part of Steve's ambivalent motivation to stay in the game.

THE CULTURE OF FOOTY

It did not take long before Steve was telling me about the abusive good-old-boys culture of the AFL—a culture of humiliation, guilt, and alienation—where the coach is king and his word is law.

S: Footy comes from an old-world perspective, it's not up-to-date in how they do things. It's an old culture, totally old culture. It's old-world. There's nothing revolutionary about footy, the same tactics and principles. Okay, some of the science might have improved, but the mentality—a lot of it hasn't improved. It's a macho game. You get all the fitness people in, and their whole preparation goes straight out the window if the coach, all of a sudden, on a whim says, "No, we're doing this, and that's what we are doing," and the fitness programmer says "Oh fuck, how am I going to rejig my whole program now because the coach wants to do this."

Steve seemed to be saying that it was difficult for anyone to question the coach on anything—sort of like a young boy finding it hard to confront his father. Steve related that an atmosphere of guilt, created by the coach, which did not seem unlike what he experienced as a junior, often left him feeling terrible about his injuries, and might have driven him to manage the injuries poorly, sometimes returning to the game too early.

S: You never play a game of footy when you're 100% fit. It's just degrees of how mentally strong you are to be able to play with an injury. At a footy club, as soon as you're injured, and you're no use to the coach, most coaches just won't talk to you; they just ignore you. They only want to know you when you actually have some benefit to them. I've been in a situation with long-term injuries, like with my groin, with my knee, and you're at a footy club; you just feel like a spare part. No one really talks to you. People come up and say, "G'day," but they don't really care. It's only a brief comment because you are not actually physically participating in what they are doing.

A "spare part"? That phrase echoes Steve's feeling of being a "commodity." Those were not the words I would have expected to hear a successful professional football player using to describe how he felt around his own club. Steve recounted that being injured was an alienating experience. He was ignored by the coach and ended up feeling victimized for something that was a normal part of the game, dealing with injury.

S: It gets to be a drain, like, you just hate going. And from the supporters, to the president, to the coach: "When are you coming back?" "Oh, indefinite, and just long-term."

They get sick of asking, and, in the end, they don't really care because they have to design their team around you not being there. So, therefore, the coach will walk past, but he won't stop and chat to you about it because he's got nothing to talk to you about.

I imagine that this alienating experience of feeling ignored due to injury increased the coercive pressures on Steve to come back too early, to train excessively, and to keep silent about niggles on other occasions. Steve commented about those coercive pressures and how he responded to them in maladaptive ways.

S: You just feel like you're useless. You want to get back because you want to be accepted, and therefore you end up doing further damage because you try and progress yourself beyond the point where you physically can do it.

Here he was, a member of a successful football club, an established player, a premiership footballer, and he felt ostracized. He just wanted to feel accepted. The emotions behind those words were palpable. The words sounded like they could just as easily have come out of the mouth of a 7-year-old boy with a broken finger, yearning to be loved by his dad. Steve's last few comments here, about the coach making him feel guilty for being injured and ostracizing him from the group, called to mind what one expert said in chapter 4 about family abuse dynamics being replicated in the coach–athlete relationship in competitive sport. The expert mentioned that an athlete can end up overtraining to try to please a demanding coach, the object of his or her attachment who has replaced Mum or Dad (see chapter 6 for more on this issue). Are the traditional practices in the sport, including the way coaches communicate with athletes, likely to put athletes such as Steve at greater risk for overtraining? From what Steve described, overtraining seems to have been a regular part of the culture and traditions in Australian Rules football, and the practices not only encouraged players to push harder in training, but also prompted maladaptive responses to injury, fatigue, and any other physical setbacks.

At this point, Steve commented that the coercive actions of the coach went beyond ignoring him and making him feel guilty. At times, the coach publicly humiliated him.

S: Coaches have said, "You're weak. You're a weak dog," or, "You're gutless. You're scared. You didn't put in," in front of your mates. And I was thinking, "I will perform better if you build me up to think I'm better than I actually am. At least I can perform at some level, but if you make me feel like crap you might get a response out of me for 5 or 10 minutes." When that emotion to prove the coach wrong is gone, I'm going to think, "Yeah, you're a prick, though, because you said this about me or that about me." We've had examples where coaches have gone and, in front of everyone, showed a highlight of some guy pulling out of a contest and just replaying and replaying it. I've been in a situation where I was being bagged about something, or being scolded about something, and you feel about this big [holds his thumb and index finger close together]. It embarrasses you so much that your teammates actually stand up for you, and it drives a wedge between the team and the coach. The coaches ball him out, and the guy's lost all his confidence, and all his teammates are trying to pick him up, and everyone hates the coach.

So, on top of anxiety and fear, players end up feeling humiliated by, and infuriated with, their coaches. The problem for athletes in professional sport is that they are tied to the team in many other ways, especially in terms of finances and contracts, and they feel they cannot openly express their anger about unfair treatment. An athlete like Steve may end up turning that anger inward on himself, looking for ways to cope with it, and respond by overtraining or coming back too early from injury, hoping to keep the coach off his back. Steve told me that sucking it up, keeping quiet about pain, putting in the extra bit of training, and pushing through injuries were expected in his sport.

S: I think it's got to do with the culture of our sport as well. I think there are a number of things. It's very much, "Do or die. Don't be a wimp. You fall, you get back up." The sport culture definitely promotes a win-at-all-costs attitude. And the whole sport definitely has no traditions that fit into taking any time off or being gentle with yourself. The "no pain,

no gain" . . . I mean, as I said, our whole culture allows no weaknesses. Nobody is comfortable with any sort of physical limitations; physical limitations are seen as excuses in our sport generally. You suck it up; you keep going, and anybody that stops for a physical reason, there's an implication that, unless you're almost dead, you know, you keep going. It's very much part of the culture; you'd better be dead before you don't go on. From a cultural point of view with sport, being tough, it's expected. Nobody thinks you're brave; it's expected. Anything less is a bit revolting and weak and snively. So, nobody says, when you play with injury, "Oh God, you're brave aren't you?" It's just, well, of course you keep going.

Damaging cultural imperatives and maxims in sport, such as "no pain, no gain," become inculcated at an early age. This phrase, along with others (e.g., "Don't be a wimp," "Don't be a girl"), are part of the sport socialization process and the transmission of the cultural values of sport. With time and repetition, these hypermasculine and emotion-silencing values may become internalized. All one has to do to see the process starting is to go to a Little League baseball game or practice and hear the coach passing these values down into the minds and bodies of 8-year-old boys and girls.

> **❝ Nobody thinks you're brave; it's expected. Anything less is a bit revolting and weak and snivelly. ❞**

As I was sitting there with Steve, the subtle sadness that had come across in his posture and expression at the beginning of the interview was now becoming painfully manifest, and at times changing form to regret, exasperation, frustration, or anger. I think I would probably feel angry, too, in Steve's situation. Furthermore, if it were not enough that the coach was being a "prick," the other staff also pushed him to disregard injuries, making Steve feel even more inadequate for having an injury.

S: You go and you report your injuries to the medical staff, and then after a while they think, "You're just whining. You just complain. You're not that bad. Look at him over there. He's got this, and he's out there training." And they pick out someone with probably a high threshold of pain and use them as a measuring stick rather than taking each case by case. It's herd mentality. "He's out there. He's got a broken rib.

He's playing. You've only got a sore hamstring or a tight back. You get out there and do it because he's worse off than you."

LIVING AND PERFORMING FOR OTHERS

Such coercive comparisons with others can be a risk factor for overtraining, and for Steve they came from many directions. Steve also talked about a sense of "the others" out there somewhere, always watching and scrutinizing his game, often unjustly criticizing his performance. He wanted to prove himself to the others. He wanted to be accepted by the others, and he did not want to let the others down. The frustration Steve expressed, however, was that many of the others, like the media, the supporters, his teammates, and even his friends and family, would judge him on his performance, without knowing that he was injured.

S: I think a lot of it is that there's a hidden pressure put on you for not wanting to let others down. A lot of the people play sports out of fear. They work so hard out of fear of letting others down, rather than enjoying the fun of the sport. You play, and you set up expectations because you play well. I'd done it to Mum and Dad and my family. When I got injured again, I just told them I felt so ashamed, and how I'd let them all down, and it was just shocking. Others expect stuff from you the whole time. Then if you're not playing well, people turn really quickly, especially when you get criticized in the paper, as happens in Melbourne, particularly with players individually. Like reporters, for no reason whatsoever. They don't know the facts behind your situation. You might have injuries. I've played with stress fractures, and they don't know that, but you're still playing because your performance is dropping. The average Joe Blow supporter doesn't know that. He hasn't read it in the paper. The reporter doesn't know that. The only people that know that are the medical staff and the coach, and they ask you to go out there and play because they need you on the ground, or need you to play, and your performance drops, and everyone's on your back. Your friends are saying, "What's wrong? Why can't you get a kick or why can't you do this?" and they don't know the pressure you're under.

Steve seemed particularly frustrated with media portrayals of his performance.

S: If I play well, I'm selfish, and I get more kicks than you. I noticed I'm in the press. I'm going to get more money. It's a tough gig. It's a balancing act. You know, the press, like it or not, you read the press, and, if you're getting shit-canned in the press, and you know you're injured, you think, "Oh, shit. The press are giving me a hard time. I've got to train harder to get back in to play, so then they can stop getting on my back." Because everyone else reads it and thinks you're a dickhead. The press gets too much power.

Everywhere Steve turned, he seemed to face scrutiny, and he gave in to the pressures. Steve played with injury; he kept quiet about his pain, hoping to please his coach, his father, the media, and the fans. I kept getting a picture in my head of a powerful and beautiful animal that had been caged. Steve did extra training to try to get better than others, hoping to feel accepted, hoping that people would not think he was a "dickhead." The paradox was that performance was both the key to getting out of that cage and, with the repeated injuries that resulted from performing, the behavior that kept him captive.

REFLECTIONS ON STEVE

As the interview was coming to an end, I paused again, observing Steve closely one final time. I could see a friendly man, a great athlete, and an intelligent, insightful football player who was open to talking about his professional career. I could also see, however, a man with regrets, whose sadness about how things could have been better might never leave him. Steve had admitted to overtraining, trying to do extra, and repeatedly trying to push through injury. Yet his overtraining behaviors seemed to be products of a relentless, abusive system; of a tough sport culture; of a money-driven industry; and of an ambitious, pushy father. The system did not seem to allow space to hear Steve, and I wondered how many other hundreds of footballers were also not being heard. How many other players also carried, or would carry, the same sadness, the regrets about abusing their bodies, about their experiences in football? Then, I thought about the thousands of kids, young footy players, aspiring to become professional footballers. For those who succeeded, would they experience such sadness when their careers ended?

Steve began learning these damaging lessons from sport and from his father (a potent combination) at an early age. By the time he reached professional football, they had become so well learned that they were almost genetic in quality. As Freud said, "The child is father to the man," and what we learn early in life shapes what we become. Children are hungry, information-seeking creatures. If the information we feed them contains messages, such as "no pain, no gain," "suck up the pain and get back out there," "footy players don't cry," and "don't be a little girl," then that is what they will learn. Like David Martin and Trisha Leahy, we all want to see sport as a place where there is joy and fun and not a school of hard knocks. We get plenty of hard knocks outside of sport. We don't need more; we need fewer. But no, we start the hard knocks early and keep knocking, and the ones who survive go on to professional sport and get beat up even more. When will the people who are supposed to care for and protect our children (parents, coaches, youth sport administrators) stand up and say, "No!"? That's a rhetorical question, but it seems the answer is, "Not right now." In Trisha Leahy's terms, many of the supposed protectors of our children keep *standing by*.

A case study, even an aggregate one, tells a tale about how an individual athlete might fit, and not fit, into his or her culture. If we have constructed a case study well, then that person's experiences may point to something universal. We think Steve's story tells us something universal about the human need for love and attention, for being held emotionally by caring others, for being accepted for who we are and not what we can do—and how, in an abusive environment, those needs can lead to self-damaging behaviors and emotional pain.

Steve's tale is about a sport where pushing excessively, keeping quiet about pain or fatigue, and playing with injury seem to be as central to the game as kicking the ball and scoring a goal. I had been immersed in Steve's sadness for the previous hour and a half, sharing his regret, his pain, his frustration, and perhaps his feelings of helplessness. When I walked away that day, I had a new and sobering perspective on professional football and on my own sadness about my competitive sport experiences. I also wanted to thank Steve for helping me think more about sport, about training, about parents, and about living with regret. I hope to have children someday, and Steve's tale has helped me renew my vow to do everything I possibly can to protect them, so that when they reach Steve's age, their joy in sport far outweighs their sadness.

CHAPTER 8

A CASE OF
OLYMPIC SEDUCTION
John's Tale

Steve's story (chapter 7) is a tale of an athlete holding onto a subtle but pervasive sadness over his sport career, and it had left me (Sean) feeling a bit melancholic. In listening to John's story, I identified with his feelings of loss as an athlete emotionally drained by his competitive sport experiences.

INTRODUCING JOHN AND HIS SEDUCTION

John had been ranked as one of the world's best in triathlon, and he was successful at multiple world events, but his three experiences with the Olympic Games showed how the seduction of the world's biggest sporting event can almost destroy some of the best athletes. Like many athletes I have talked with, John started out with early successes and a rapid rise in his sport; he also began his career with high expectations from himself and others. The early potential, the promise of success, and the drive to be best in the world led to fantasies about the possibility of triumphing at the pinnacle of sport, the Olympic Games. That dream, that possibility, acted on John in ways that seduced him. He was driven by an intoxicating desire to succeed in his sport, yet he was blinded in his pursuit and ended up behaving in ways that ensured his dream would not come true.

John had a huge internal drive to succeed and good knowledge of what worked for him in terms of training and recovery. Unfortunately, he got it wrong each time he tried to achieve his ultimate dream of Olympic gold. At his first Games, in 1992, John thought he had done everything right; he was going to take the world by storm. He was going to surprise everyone. He was flying, but he ended up sick, fatigued, and unable to perform. At his second Games, in 1996, John did not want anything to get in his way. He had been there before; he thought he knew what he had to do to be at his best, but injury, and poor responses to injury during the lead-up to the Games, cast him as (in his eyes) a repeat failure. He was devastated about once again being unable to compete anywhere near his capacity. At his third Games, in his home country, John wanted to get it *perfect*. It was to be his last Olympics. This last time around he had more experience and tremendous success at the world level and knew the formula for winning. But he believed he had to find something extra, something superhuman, and ended up driving himself into a state of overtraining syndrome, not even giving himself a chance to compete. As the interview got under way, I could not help feeling a connection with John. He had been devastated, disillusioned by the promise of Olympic glory. It felt like something that I knew well, and I was eager to help him tell his story.

OLYMPIC GOLD AS LOVE OBJECT

We began talking about what performing at the Olympics means to athletes. John said that he had dreamed of representing his country at the Olympics for much of his life. He talked about how he ate, drank, slept, and breathed Olympic glory.

Life would be complete if he could show the world, on sport's grandest stage, that he could be the best. The image was beautiful: standing there on the podium, tears of joy running down his cheeks for his family, friends, teammates, coaches, competitors, fellow citizens, and people of the world to see. The dream of Olympic glory was undeniably seductive. Unfortunately, John responded like a bowled-over lover; seduced and blinded to any objectivity, he pursued the dream too ardently, too hungrily.

Wanting so much for the dream to come true, John ended up seriously damaging himself three times in vain attempts to do more than his body could handle. The fantasized tears of joy ended up being tears of pain.

Pursuing the Olympic Lover: Act I

On several occasions, John talked about the Olympics with an excited tone in his voice. Recalling his lead-up to the 1992 games, John said he felt overwhelmed but positive and motivated by the prospects of success and celebrity.

John (J): It was the Olympics! Holy shit! They're finally here after all the waiting! This is *the* main goal. This is it in front of you, and all of a sudden you've ticked off all the boxes. Now it's the Olympics that you've always been waiting for. You've always dreamed that you're going to be the best, and you win the gold and, you know, become a household name and be a hero.

There it was, Olympic gold, and the promise of eternal love and acceptance from everyone—a household name, a hero! I had a sense of what John was talking about, having pictured myself many times with an Olympic gold medal hanging around my neck. The picture could be intoxicating, the type of image that can push athletes such as John into the dangerous realms of overtraining (OT). He developed many of the symptoms of overtraining syndrome (e.g., immunosuppression, persistent debilitating fatigue). After his first Olympic disappointment, however, John's overwhelming, dreamlike motivation began to shift to more of a desperation permeated with fears of failure.

> **" You've always dreamed that you're going to be the best, and you win the gold and, you know, become a household name and be a hero. "**

Pursuing the Olympic Lover: Act II

J: The closer I got to Olympic trials, and the less my back was responding, and the less I was doing, the more I'd resigned myself to, "Okay. Look. I've got to just get there. I've got to get there. If I don't make these Olympic Games . . . No! That is not an option!" I couldn't deal with that prospect, no way. It was the biggest thing that I ever wanted to do, and I had to succeed. I mean failure was not an option.

During his preparation for this second attempt to compete at the Olympics, John sustained an ankle injury but still had enough time to respond to it, rehabilitate, and compete. I wondered how his perception that his dreams might once again slip away affected his response to that injury. It made me think of what several experts in chapter 4 said about how athletes at risk for overtraining might react desperately, trying to overdo everything to get back to their sport, often returning too early from the rehabilitation process. I wondered: If John had managed his injury differently at the time when it first occurred, would he have been in a better position to compete when he finally got to the Games? I also wondered if John had been blocking out the early signs of the injury while training because he did not want to acknowledge that there could be anything wrong with him. He had said that failure was not an option, a profound statement alluding to the pathology underlying John's pursuit of Olympic glory. In shutting out the possibility of failure, which seemed to be unfathomably anxiety provoking, John created a pattern of defensive responses that were more likely to produce the most feared outcomes than prevent them. I would have to wait to see how John's story unfolded to understand the complexities of this behavior that led, ultimately, to his unfortunate experiences.

Pursuing the Olympic Lover: Act III

In talking about his last attempt at the Olympic triathlon, John again voiced desperation. By this third time around, however, he had accumulated successes at the international level in his sport. He had learned what it took to be one of the best in the world, winning World Cups and World Championships. The Olympics, however, held a mystique, a seductive pull that overpowered his logic, knowledge, and experience.

J: I knew I was getting close to the end of my career, and the Olympics were the last thing I wanted to do. I mean, this is the biggest thing I've ever wanted to do! You know, I was a realistic chance to win the Olympics.

The possible double meaning of the Olympics being "the last thing" John wanted to do is striking. John said that the Olympics were the last thing he wanted to do before retiring, but I had to wonder if he might also have been saying, albeit unconsciously, that the Olympics were something he would never want to do again, especially given his previous two Olympic experiences. Regarding this third Olympic attempt, John described the pressures to perform and his fears of failure as bearing down on him like a predator hunting its helpless prey. He really began to question his ability, but he could not turn away; he could not give up the dream.

J: What if I can't do it? When you are on your way up, you're like, "What if I can? Yeah! What if I can?" When you're young and when you're coming up, you've got the pressure behind you, and under you, lifting you up, whereas when you get to the top, it's like the pressure is coming on top of you and bearing down on you, because you're just like, "What if I don't live up to the expectation? Oh! What if I do that?" You put pressure on yourself more than anything else because you want it so badly. So ultimately you are controlling the pressure, but most people let the pressure control them, don't they? I had to go for it. I had no choice.

From John's descriptions here, I could see that he appeared motivated to push himself. He admitted to putting a lot of pressure on himself, but the contradiction between his statements—"You are controlling the pressure" versus "I had no choice"—suggests that he probably felt many other pressures beyond his own internal drive. As in his first Olympics, he overtrained and "crashed" with serious illness and fatigue. His story brought me back to one of the first questions underlying this whole project: Where does this drive come from for athletes to do extra—to neglect recovery, injury, and illness—despite knowledge and experiences that should have taught them to pursue balance

in their training and recovery? Like the promise of a professional contract, an Olympic gold medal seemed to be a huge motivator. Similar to the professional athlete, the Olympic dreamer also had, along with his internal drives, a whole system encouraging and colluding with him in his relentless pursuit of success.

EARLY SUCCESSES

Looking more closely at what was driving John, I asked him about his early experiences in triathlon. Not surprisingly, like many before him, John was lured into the world of competitive sport by showing promise at an early age. He progressed quickly, accelerating through the ranks of junior and then senior triathlon. Success can be seductive; John loved his sport, loved success, and was thrilled to find that it came to him easily.

J: Heading towards the end of high school, I'd done really well in athletics in interclub and especially when I was in year 11 and 12 [junior and senior years in high school], broke a few interschool records and got recognition for being a good athlete. I did my first big triathlon at the Victorian Championships, and I won Under 20. I was still a junior, Under 20, and I won the Victorian Title. I won the national Under 20 title, and from there I never questioned it again, and I didn't look back. I thought, well, I've sort of got some instant success, and then I was hooked into it. I loved it!

John was good at his sport; people applauded his success, and, possibly, he had found a way to feel loved. Perhaps John had found a source from which to fill a void, a way to compensate for feelings of insecurity or inadequacy. Perhaps, for John, success became like a drug; it felt really good at first, but then he needed more and more of it to get that same feeling. These thoughts were my tentative speculations stemming from my psychodynamic training, along with the knowledge that John overzealously pursued Olympic fame and adulation ("the world will know I'm the best") three times and crashed in all of his attempts. With the taste of early success, and the recognition of his ability, the first act of John's seduction was complete. All he needed now was a little prompting from the right quarters, and the Olympic dream would take firm possession of his psyche. It didn't take long, as the media and esteemed coaches were soon encouraging John to push on.

J: I got interviewed by a few press people in the newspaper, and they asked me if I'd ever do this for 1992. I think it was just after the [1988] Olympics, and they asked me, "Oh, are you going to continue with triathlon? You broke the record!" I sort of said, "Yeah, I could go to the Olympics." Now, why I said that, I still don't know, really. So I said that as a throwaway line, and I think I might have mentioned that [a particular triathlete] was a bit of a childhood idol. My coach [who was the top coach in the country for the sport] was on a plane to Europe, picked up the newspaper on the way out, at the airport, and read it on the plane. He came back and told me I had a great amount of potential, thought I'd be perfect for Olympic triathlon. He thought I had a real future in triathlon and asked me for a 2-year commitment.

John did give his time, his commitment, and his heart to the sport of triathlon, as his coach had asked him to do. As with Steve, the footballer portrayed in chapter 7, John also experienced tremendous success in his sport. Outside of the Olympics, John would rise to be one of the best in the world.

KNOWLEDGE VERSUS THE HUMAN HEART

J: I had 10 years of professional triathlon, won world titles, five medals at World Championships, so you know I've been at the top level. I knew what I had to do to get it right.

John said that he knew how to get it right. Nonetheless, as the successes accumulated in his burgeoning career, John also began to experience increasing pressures as well. He felt the pressure from within, the personal drive to live up to his potential, and he felt the pressure from outside, the expectations of others, family, friends, the public, the media, and his country. It seemed to John that the one way to resolve all this pressure, to live up to expectations, was to capture the holy grail of sport, to triumph at the Olympic Games.

Looking at John's tremendous success at the world level, it seemed out of character that he repeatedly came up short at the Olympics, because he seemed to know how to get it right at other times. John even talked about having the knowl-

edge about training and recovery to do what was best for him.

J: I didn't have a problem trying to do too much because I knew I'd worked out what was the optimal amount for me. I didn't need to do more volume, you know, and that's the beauty of it. It's finding out what's optimal for you, because everybody is different in what they can handle and what works for them, and it's really that simple. I was really pretty good, generally speaking, about injuries and illnesses, and I did take it in my stride and go, "Okay, well, I've got this problem. What do I do to overcome it? How quickly can I overcome it? How quickly can I get over it?" I mean, I'd be pretty good at taking a rest if I needed it to get over something.

I took note of John's perception of himself as someone who did not have a problem with trying to do more than he should, despite that he had done exactly that in trying to prepare for all three of his Olympic campaigns. Once again, I was thinking about what the experts in chapter 4 said about the motivational elements of something like the Olympics, that it prompts coaches and athletes alike to look for something extra, to change what has worked in the past. Although John did get it right a number of times at major international events, I had the suspicion that I would hear about some behaviors and attitudes, later in the interview, that would contradict his perception that he did not have the problem of trying to do too much, or that he was pretty good about responding to injuries and illnesses. Not surprisingly, in the following discussion, he dropped the first hints that he might push the limits of his training and recovery.

J: There were times when, you know, due to scheduling and races coming up, if I had an injury, I'd like to find out what it was, and I'd like to find out if I kept pushing through it, if I'd make it worse, or it'd just stay the same. If it would just stay the same, I could push through it for a period of time until World Championships were over or something and then take a rest. I'd do it that way. I mean, I'm pretty logical and pretty analytical about things.

In the preceding quote, John was maintaining that he was generally good about handling illness and injury, but I wondered: If he were so good at dealing with injury and illness, why had he got it wrong at three Olympics? I asked John about these contradictions in his thoughts and behaviors, and he pointed out that he ended up changing his mind-set when it came to the Olympics. He said that in his first Olympics he was feeling so great before he got there that he had reached a point where he was feeling invincible.

J: Before I got to the Olympics, the lead-up, everything I did was golden. I was just rocking! The PBs [personal bests] were coming out everywhere, and I was just flying! Went to Spain, continued that form just so keyed up. I thought, "I'm going to take the world by storm!" You know, it didn't matter; no one would know me, and I had that advantage, and no one would be worried about me. I'd just come out of nowhere and just take them all apart! The form that I was showing leading up, there was nothing to lose. It was my first senior team, and I just thought I'm just going to go and tear it apart!

Unfortunately, what he did was tear himself apart, physically and emotionally. He reached a great peak in his training, then went on to push to another peak. He went to altitude and trained even harder. He so depleted his resources that he compromised his immune system and fell ill. That extreme level of motivation to perform at the Olympics, the belief that one will succeed, and the giddiness of reaching excellent physical form prior to the Games can push athletes and coaches to do too much. I thought a bit more about this sense of invincibility, wondering where it comes from, how the Olympics drive athletes to delusions of grandeur and feelings of godlike stature. Perhaps feeling invincible results from getting ahead of oneself, fantasizing about the perfect outcome before it is attained—in this case, the reward of absolute love and approval for being an Olympic champion. I remembered one of the experts from chapter 4 saying that he was most concerned about athletes being at risk for overtraining when they are in peak form. He explained that athletes who have reached new peaks, or achieved recent PBs, are at risk for two major reasons. First, on the physical side, to reach the peak, athletes have probably balanced their training and recovery just right; therefore, if they do more, they may push themselves into overtrain-

> **" ...and then why the hell we decided to up my training and try and reach another peak? I have no idea. In hindsight, it was just stupid. "**

ing. Second, on the psychological side, the thrill of peak form, or new PBs, is so motivating that athletes and their coaches end up thinking, "If we can do this well with this much training, imagine how well we could do with a little more!" Listening to John's story about preparing for his first Olympics, it seemed that both he and his coach got carried away with his excellent form. Sometimes, solid knowledge of how to do it right cannot compete with the desires of the human heart.

J: I had qualified, much to everybody's amazement, including mine, had a bit of a blinder [a great race] of a race, and then I dominated the rest of the selection races. But then we did something fairly foolish leading up to Barcelona. We went up to altitude, and I was in really, really good form, and I, foolishly, with Coach, I guess we made a big mistake. We did too many hours at altitude. I had already peaked, and we were trying for a bit more training. We just pushed it too far. I was already in fantastic form. I made this huge leap and just killed the other guys, and then why the hell we decided to up my training and try and reach another peak? I have no idea. In hindsight, it was just stupid. Why, when I'd already made such a quantum leap, had we tried to go that extra bit? I was really susceptible to getting whatever virus was floating around as soon as we got to the Olympic Village with people from all over the world. I just lost everything, all the strength, and there was nothing I could do. It was a virus that went through a few different teams, and I got smashed by it. I was in the best form I've ever been in when I went down.

John admitted that he and his coach made a mistake by trying to reach a new peak when he was already in such good form. Despite being ill in the week before his event, however, John proceeded to compete at the Olympics, perhaps setting up a behavior pattern of competing with physical setbacks that might have gone against better judgment.

J: I raced, but I probably shouldn't have. Yeah, I raced. Stupid! I ended up getting a Mars Bar off a photographer at the last 5-kilometer mark to see me over the line, but I was right off the pace. Scott had a really good Olympics, and I had been knocking him off convincingly. So, he took all the glory, and I was sick, and that was my first Olympics.

As I looked over at John during the interview, describing his first Olympic experience, he appeared shaken, sad, frustrated, and even pissed off that he had gotten it all wrong. He could not help expressing his regret to me.

J: If only I had looked back and seen where I'd come from and what sort of form I was in and gone, "Well, look, you've made this huge leap! It's a stressful time. Just take it really easy leading up into the Village. There's no point in doing big hours." Why do people go do these stupid things? Why can't they just stand back and think? They sort of know they're doing it, but they do it anyway. I knew, and I did it, and three times I blew it.

I empathized with those feelings of regret because I had gone through a similar experience in my attempt to make the Olympics, but I still wondered why John had mishandled the Olympics twice more after that first bad experience. In the next few minutes of the interview, John really opened up on an emotional level, letting me glimpse the depth of his passion for his sport and, perhaps, the depth of his possibly pathological need to succeed. On top of being sick and fatigued during his first Olympics, John discussed the humiliation he suffered for being one of the last athletes to get out of the water on the swimming leg, a discipline he was supposed to dominate, and then being one of the last athletes across the finish line.

J: I was about 5 meters down on one of the last guys. I was so frustrated. I got out of the water almost last. I was screaming inside, just yelling at myself. Oh! It was shocking, and I just felt so humiliated and so embarrassed in front of the whole world. As the race came to an end, I was just running for my life, basically. It was the weirdest experience; it was like one

of those dreams where you can't run away from the guy that's chasing you, or you're trying to catch something, and you're just running on the spot. It was almost like that feeling. At the end, I crossed the line, and I was a mixture of anger and humiliation. As you cross the line at the Olympics, there are 58 TV cameras, just all round the end, so you run across the line, and then all you see is cameras. I just sort of put my head down, and I just walked past the guy who won the race, and I said [to himself], "That was fucking bullshit!" I walked past, and I had my head hung low and, like I said, I was so shattered. I didn't know what to do. I was so shattered. I was sick, but racing the way I did just piled on the humiliation. It was like I was being crucified in front of the whole world.

THE DEVASTATION OF SHATTERED DREAMS

It sounded like performing poorly at the Olympics was more than just disappointing for John. His words—"shattered," "embarrassed," "humiliation," "crucified"—conjured up images of John being tortured from within, of the insatiable hunger of his narcissistic psyche driving his self-loathing for failed Olympic dreams. John's experience of humiliation speaks to how important showing the world his talents was for him, how much he wanted that external adulation. From delusions of grandeur, of achieving godlike status, John had descended to images of being a crucified martyr. He had said that he held a lot of his emotions inside, but he told me that at one point, with his girlfriend, he also let it all come flooding out.

> **" I was so shattered. I was sick, but racing the way I did just piled on the humiliation. It was like I was being crucified in front of the whole world. "**

J: One night we were lying in bed, and we were talking about stuff, and I don't think I had cried in front of her before. Oh, this night I cried my eyes out, and she was like, "What's wrong? What's wrong?" And I was like a baby, and I probably cried for an hour straight. I couldn't stop. It was like I didn't care that she was there. I didn't care that I was embarrassed. I didn't care about anything. I had to lie in the bed and cry and cry and cry. I've never cried that much, ever, and I really didn't even talk. I didn't even tell her why I was crying, I don't think. I just remember crying, and it lasted all night, but it was such a cleansing feeling. It felt good to cry, and so I just rode it all the way, I rode it all the way out of my system. I mean there's still a deep scar from that somewhere, but I can deal with it now, and I've dealt with it.

John's recollection, here, of his outpouring of emotion surrounding his first Olympic experience, made me think about how the emotional and physical trauma of competition, and the deep scars that get left behind, have effects on subsequent competitive experiences. Is an athlete who is affected so profoundly by a humiliating sport experience more likely than others to push him- or herself too hard in the future? Was John's emotional response a hint, a warning sign of things to come? Were most of his following 8 years of training and continued pursuit of Olympic gold about trying to heal that deep scar? Was it John's attachment to fixing himself through sport that drove him to get it wrong in two more Olympic Games? John did get it right several times at the world level, but I had the feeling, despite John's comments that he had worked out what was optimal for him, that he might have got it wrong at other times besides his failed Olympic attempts. I thought that when it came to overcoming his regrets, and learning from his mistakes, John actually had not "dealt with it."

REPEATING PATTERNS

In his pursuit of Olympic glory, John learned much. He learned about discipline and commitment. He learned about sacrifice. He learned to push himself. He learned to persevere in the face of setbacks in the form of fatigue, pain, injury, and illness. He also learned about isolation, fear, and disappointment. Ultimately, he learned about failure. John admitted to learning about many things too late. He acknowledged adopting patterns of behavior that led to destructive results for his body and his performance at his three consecutive Olympics. As our interview moved into discussions of John's experiences with illness and injury outside of the Olympic periods, I began to see that perhaps he was not as balanced as he had earlier suggested. With the Olympic dream branded onto John's mind for 15 years, he had felt compelled to block out injury,

illness, and fatigue, or, at times, to push excessively after already having made leaps in performance quality. In the following comments, John highlighted several times when he got sick, but chose to continue training or competing.

J: I went over to World Student Games and went overseas to Europe as usual. I was in France, and I was quite crook [ill] in France. I did race, but I wasn't healthy. Quite a few people got sick, but I got very sick, and then we went straight on training anyway. Then we went to World Juniors, and I ended up racing, actually. I think I started to get better, but then I went downhill again, and I went to see the doctor, and he did tests on me and said, "No. You're not starting." That was sort of the start of my illnesses. After that I qualified in the first World Cup, and then 5 days later, I don't know why, I got sick again. It was pretty bad. I got really bad migraines, headaches, and I'd just lie on the floor thinking I was dying. All the doctor could say was that, yeah, I've got some virus. It happens to quite a few triathletes every season, so I should write this one off, go home, and come back next year, but I didn't really know what was wrong, so I kept going. I sort of got better, but my pulse was still messed up. We used the old test, lie down and stand up and watch your recovery rate, and it was just stupid, and when I went out training long slow distances [hours and hours of jogging], it just went through the roof, but I didn't actually feel that bad. So it was the end of June, and I went down for a World Cup in Austria, and on the first hill of the run, I just went so lactic, the worst race of my life. I should have pulled out, but I didn't because I'd never pulled out of a race before.

The pattern evident in John's discussion of illness was that he would continue to compete despite feeling sick, and sometimes despite medical advice to take time off. Somewhere in John's development as an athlete, he had learned to keep pushing despite having an illness, something that the experts in chapter 4 pointed out as a high-risk behavior for overtraining outcomes. I had the sense that John was going to reveal more about where he had picked up his compulsions to push through illness. I was soon to learn that John also had stories to tell about injury.

J: I often wonder where I could have gone if I would have had more time to train and

. . . been smarter in what I did. I had quite a few injuries too, though. . . . I had problems with my ankle from overuse. I had some problem with my low back that I never really managed to isolate. Tendonitis on the left knee, which now I've got problems with, plus the world's dodgiest ankles and skinniest Achilles tendons you've ever seen.

On a number of occasions when I asked John for more detail about his injuries, he surprised me. In his description of the lead-up to his first Olympics, John had mentioned that he was flying, was in peak physical form, and, although he had done too many hours at altitude, he had said that nothing was wrong with him until he got sick at the Olympic Village. Nonetheless, when we got to talking about injury, John pointed out that he had experienced problems with tendonitis in his left knee prior to his first Olympics. Memory is a beautiful thing, and with the help of the human heart's desires, it forgets a lot, and I sensed I was going to hear about several more instances of John's forgetfulness.

J: Going back a bit, just before Barcelona, I got tendonitis in my left knee. So this tendonitis got to the point where it was really giving me the shits before we went to Barcelona. I hated the thought of getting the wear-and-tear kind of injury so early, and that was sort of playing on my mind. I thought, "I've got chronic injury, and I'm only 22."

I saw John's omission of this injury in the first telling of his Olympic experience as support for my growing sense that he might have exhibited behavior patterns throughout his career that put him at greater risk for overtraining and injury. The behavior patterns seemed to revolve around keeping silent about injury and illness, trying to block them out in the early stages of onset, and, generally, trying to make everything sound great when talking about his health. Or, it may have been that early in his career he did not understand the processes of overtraining, but that possibility does not explain his later behaviors when he certainly knew what was good for him. His denial or omission of illness and injury was a pattern that John repeated throughout his career, and repeated again with me in our interview, at least until he began to feel more comfortable talking to me. With an interest in understanding John's attitudes toward injury, I asked him the following question: "You didn't feel that knee injury coming on?" His answer

seemed to support my thoughts about his trying to block out the injuries: "A little bit, but I thought, 'That can't be happening, not this early!'"

With respect to other injuries, it turned out that John had to miss several National Championships, World Cup events, and World Championships over the years because of injury. He gave me a summary of a few of his injury experiences.

J: My back became a real issue, and as a result I got to the Nationals and couldn't really get into the start, and my back seized up. Then after that big winter of 1993, I got to Nationals that summer, and I screwed my right ankle, ligament strain, screwed it and so subsequently pulled out. Won the National Championships at the start of 1995 and went to Europe for an invitational. The best got invited to that, and again I was in really good shape, really great. Then on the bike, I strained an abdominal muscle, like a psoas, so that put paid to [wiped out] my Europe campaign. Lying 4th and had to withdraw, so that really pissed me off.

As John's injury history unfolded, it was remarkable how many injuries he had endured and how often he had mismanaged them. He seemed to be engaging in exactly the sort of OT behaviors with regard to injuries that the experts described (e.g., ignoring pain, not reporting discomfort until it is too late). On another occasion, John talked about continuing to train and compete despite what seemed to be considerable pain. He also admitted to keeping quiet about his pain in front of his coach, at least until it became too debilitating to mask.

J: So I'm in [the UK]. I went to see this guy just to get a rub on the back. He was doing this stuff, then he sort of started contorting me. He did some rotations and sort of clunked me, and I thought, "Yeah, that killed." He said, "Look. You'll feel terrible for the rest of today and tomorrow. Neurologically, you'll feel crippled. You won't be able to do anything. I would recommend you didn't train tomorrow, but after that you'll feel roses." The neural stuff had settled down, but, all of a sudden, where I'd had some pain I had a real spot, like a stab right in my spine. Oh, it would take my breath away to sort of rotate on it. Anyway, I kept it under my lid a bit, didn't tell Coach, and we traveled around, did some more comps, and it was starting to really upset me. I didn't want to tell Coach until it got to the point where I said, "Look, that chiropractor I saw, whatever he did when he clunked my back, he's done something, and it's messed it up, big time." I was starting to get to the point where I couldn't swim, couldn't run properly, couldn't do any rotation, couldn't do any impacting, which basically is everything. Physios [physical therapists] were talking: "Look, if it doesn't settle down, we can just give you an injection." So, I was like, "Okay. That'll be good." So, treatment, treatment, treatment. Nothing was working on the treatment side. I was still training, fully expecting to go and do World Champs. It'll be fine. Last resort, get a jab, no worries.

"Get a jab." That was something that I heard from a number of athletes, especially from Steve the footballer (chapter 7). John was trying to hide his injury and not communicate about it to his coach. To parallel this type of concealing reaction to injury, John had a system behind him that would help him deal with his injury by numbing it with cortisone injections. As with Steve, I was getting insights about the pressures (to train and compete in pain, to use drugs to mask pain in order to compete, and to risk more serious injury while competing when already injured) within John and within the sporting environment that might have driven him to push through injury. When it came to John's second Olympics, he responded questionably to an injury that preceded the trials, and that response seemed to have affected his ability to perform to his potential when he got to the Games.

J: I just rolled my ankle over in training and had fallen like that, heard it crack. I was, "Oh my God!" I thought it was broken. I heard this crunch, and I just lay there screaming. You know, you say that your life flashes in front of you. My Olympic dreams flashed in front of me. I just thought, "There goes the Olympics, straight away. Gone." I'm lying on the bed, and I didn't want to look down. I was expecting to see it hanging off at an angle, and I was screaming, and swearing, and cursing, and the coach was running around. I got the courage to look down, and it was like, "Okay. It's attached." I tried to move it, and I could move it, and it was killing me anyway. I went and I put ice straight on it. I set my alarm for

every hour, had a bucket of ice next to the bed, and every hour got up, put my foot in the bucket for 15 minutes. Go to sleep, hour later, and that was for the first 48 hours; every hour for 48 hours I had ice on it, just knowing that time was my enemy at that stage. I was overloading the rehab. Everything I was doing had to be accelerated. I didn't want to waste time. I didn't have time to get it right and then start building it up. It was just a matter of, "Get it right as best you can, because you've got to." I didn't want to go to the Olympics and *just* participate. That lead-up period was supposed to be where you start to crunch, like in all the years gone by, that period where you lead up, you start clicking everything together. So that was eating into my period where I was supposed to be getting speed up and getting all the things to click and get ready. I ended up doing the best I could rehabbing it. I ended up competing at the trials with it taped. I won the trials in not a very good time. I got through the trials okay. My ankle was okay. How stupid was that, though? The ankle sprain was 11 weeks out, and I panicked when I shouldn't have because my normal prep is only 8 weeks anyway, and I still had 8 weeks once it got healed.

> **" *I was overloading the rehab...I didn't have time to get it right and then start building it up.* "**

John was feeling desperate when he hurt his ankle; his Olympic dreams flashed before his eyes. In retrospect, he acknowledged that he had adequate time after the injury healed to prepare properly for these Olympics, but at the time of the injury he could not see that clearly. That desperation about the possibility of missing another Olympics led John to panic. He might have to forego another chance to fulfill his intrapsychic needs and to impress and win the love and affection of the triathlon world for his athletic achievement. As he said when he made his first Olympic team:

J: Qualified for the Olympics . . . so was thrilled with that. Deep down we knew that we could do it and, when we pulled it off, it was great and the whole [triathlon] world couldn't believe it. They were just like, "Where did this kid come from? How did he do it?" People were really impressed and all the rest of it.

The perception of not having enough time, and John's sense that the injury was eating away at his preparation time, seemed to motivate him to do more than his body could handle in terms of training and rehabilitation. He did not allow sufficient time for the ankle to heal, pushed the injury too hard in both rehabilitation and training, and ended up unfit for the Olympics.

In his first two Olympics, John talked mostly about bad luck with getting sick or injured. He claimed that he was not the kind of athlete who went overboard in terms of physical training, stating, "I never push myself in terms of doing the physical side. I push myself on the technical side." Yet now John acknowledged that, eventually, he did push himself physically, in the lead-up to his final Olympics.

J: I sort of got to the stage where it was like, "I'm just going to train!" That was the important thing, and if I hadn't had good enough sleep, or I hadn't eaten well enough, it didn't matter. I was training, you know, but I went overboard. I did too much. I think, when you really get classic chronic overtraining syndrome, you're on a fine line, and you don't realize it, and then you just go bang off the edge, which is totally what I did. I went from here, getting really fit, to crashing. Walking down the beach 100 meters from my front door was an effort. I had to rest before I walked back.

STRUGGLING TO UNDERSTAND

I was getting a more developed picture of John and his overtraining behaviors. He had originally made it sound like he had had a lot of bad luck during his Olympic campaigns. He had said that he was good with injuries and illness. He had said that he knew what was optimal training for him and that he did not push himself too hard. It had become clear to me that John was not just a victim of bad luck. John contradicted his original statements about being good with injury and illness and not pushing himself too hard. He was struggling to understand and explain himself.

Looking back on his last Olympic campaign, John acknowledged that he had been fully drawn in by the Olympic dream once more. The seductive pull of gold medals had once again obscured his objectivity and set in motion his single-minded pursuit of perfection. John got trapped in feeling that he had to do something extra, something special, when it came to the Olympics.

J: Oh, I just tried too hard. Yes, just didn't stick to the normal formula, went for that bit extra, you know, did things a different way, panicked when something went wrong. I think the mistake people make is they just look for that bit extra, or try for perfection, or try for too much. I know for me, when I overtrained, at the end, leading up to Olympic trials, I basically tried to get my preparation perfect. I'd spent 10 years; I'd been top 6 in the world. I'd been first, fifth, second, first, second, and I'd done every one of them on 8 weeks or less prep, and none of them had been perfect. For Olympic trials I tried to have a 3-month prep. What was I thinking? I tried to do it perfectly. I tried. I'm like, "Okay, for the first time in my career, I'm going to get it right. I'm going to. I'm going to have the big prep. I'm going to put in the big work." It was stupid, because it wasn't what had worked in the past. I mean I was probably better off to stuff around [let go of] everything up until the last 6 weeks and then put in the big work from then on.

John was trying to get it perfect (he mentions the word *perfect* four times in the preceding comment), but ended up getting it badly wrong. He erred by looking for something extra, something special, something different from what had worked for him in the past. In this final attempt at the Olympics, armored with World Championship successes, equipped with knowledge and experience from previous Olympics, he still wanted to change anything he could to get it *perfect*. It was to be his last dance, and he surrendered himself one more time to the Olympic seduction. He did not talk so much about his feelings of invincibility or his delusions of grandeur in this last Olympic attempt, but he still sounded desperate to fill a void. Perhaps John could show the world how great he was and be loved and showered with adulation, but only if he could be *perfect*. He even changed coaches for his final Olympics, and he described the change as a desperate attempt to maintain a sense of control over his training.

J: I still know that if I'd been with my first coach I would have gone to the Olympics, but I made the choice that I didn't want to do it that way, and I only wanted myself to blame, and I wasn't happy in the training environment with him. I had another coach who was really good leading up to the last Olympics, but I didn't allow him to have control, and he was too scared to do it. Certainly, everybody expected me to do well at the Olympics, but the pressure and expectations from myself were probably higher than anybody else.

It sounded to me like John was at that point flailing in his efforts to have a sense of control over his Olympic campaign. Over the years, John's Olympic dream had been transformed into more of an Olympic nightmare. He had been maimed, humiliated, and crucified. In his first Olympics, it had been all about the team of himself and his coach; it had been about a shared dream of Olympic gold. In his last Olympics, John had not wanted to trust another coach who might let him down again. John had felt that his coach had failed him in the first two Olympics, and for his third he had not wanted to blame anyone but himself. In the end, he seemed to get what he wanted.

In response to John's comments about putting pressure on himself, I wanted to understand more about how his character might have influenced his overtraining behaviors and attitudes. When I asked John about what he saw as some of the driving factors behind his behaviors, attitudes, and unfortunate overtraining experiences, he began by telling me something about his own personality. John acknowledged his strong internal drives. He called himself a perfectionist and noted that, at times, he was obsessive in his approach to training.

J: I'm totally a perfectionist, and I'm totally obsessive, and I'm totally anal, and all that sort of stuff, but, yes, I'm good. I think I can say *better* than I can do. Probably always have been like that because I've always been able to think very logically. I mean, I can be superhard on myself, and I think at the top level you are superhard on yourself. Certainly, athletes are really hard on themselves. I tend to be like most athletes who do what I've done, or what you've done. We're pretty driven in anything we do. I think that's a really good quality I have, and I think being hard on yourself drives you to be one of the best, but I think you need external people to moderate that. Like, it's okay if the athletes are really hard on themselves, as long as they've got somebody watching out for them.

The mass of contradictions continued. Just after telling me how stupid he felt for trying to get his last Olympic preparation perfect, John defended himself by saying that it is good to be a perfectionist, to be self-critical, to be superhard on oneself. John did note that it is important to have other people to moderate a self-critical athlete, but when it came to his last Olympics, John went against his own advice and shut out his coaches. This ongoing string of contradictions seemed to exemplify John's career, and left me thinking about how competitive sport creates an environment where balance itself is in contradiction with the demands of competitive sport.

Reflecting on balance, I thought about my own experiences in rowing. I think John must have sensed my identification with his situation, because he suggested on a few occasions that he and I might be similar in our attitudes or behaviors, as in this comment (italics indicate my emphasis):

J: I was always my worst, and still am my worst, critic. I'm always harder on myself than anyone else. I don't know whether that's the same with people like me or not, or I don't know whether it's the same with *you*. I guess I'm not maybe as laid-back as some people. There's never enough time, I guess. I am a bit that way with everything, like I want it to happen tomorrow. I never watch telly [television]. I've always—*you're probably the same*—but bang, bang, bang, 50 million things on the go, rushing here, rushing there, 5 minutes late for everything. I've never thought of myself as the best in the world. I don't think I'm some kind of sick person. I'm just me, and I'm an insecure person *like the rest of us*. I'm just a normal person, right?

Asking if things were the same with me, saying I was probably the same as him, and then saying he was an insecure person like the rest of us, John seemed to be reaching out to me. It seemed like he wanted me to normalize his situation and say everything was okay. He had taken the blame for his failed Olympic dreams, but he wanted to share his confusion and hurt. John seemed to want me to validate his normality; perhaps he wanted me to tell him that he was deserving of love, like any normal person. The funny thing for me was that I felt full of contradictions in how I responded internally to John's comments. I thought, "Yeah,

> **❝ ...it's okay if the athletes are really hard on themselves, as long as they've got somebody watching out for them. ❞**

I suppose we are the same in some ways, but no, I am different. Perhaps I could be hard on myself when I was training, but I was pretty laid-back sometimes. Yeah, I like to have a lot of things on the go, but *surely* I was more balanced than John, wasn't I? Anyway, I am not an insecure person!" All of a sudden it struck me. Like John, I had my own mass of contradictions that were part of a behavior pattern, a pattern that allowed me to keep pushing myself despite knowledge and awareness about overtraining (more on this topic in my own story in chapter 10). Thinking about my own situation and John's together, and realizing that we both had strong drives, moderated by reasonable knowledge about training and recovery, I wondered again what experiences or relationships with others might have reinforced John's contradictory behaviors, thoughts, and emotions.

INTERPERSONAL INFLUENCES DRIVING OVERTRAINING

Although John had a strong internal drive to achieve Olympic stardom, there seemed to be a number of people behind him who were colluding in his self-destructive pursuit of Olympic gold, who trumpeted the seductive myths of Olympic success, and who taught and reinforced many of his maladaptive behaviors. The most influential people in John's athletic career seemed to be his coaches, especially the first coach who helped develop John's skills as an Olympic contender in triathlon. John talked a lot about his first coach; he respected the coach's knowledge, drive, and skills in pushing him to his limits. He made comments, however, about his coach's attitudes toward training and recovery that revealed where he might have learned his maladaptive behaviors. From the beginning, John said that this first coach had to show him the necessity of doing more work than he ever would have imagined himself capable of handling.

J: Yeah, I committed. I started April Fools' Day in 1990, actually, and started running, and it really was a wake-up call. I couldn't believe the training that he expected of me, and even then, looking back, he really

> **" Well, I probably didn't tell Coach about the injuries…he couldn't deal with injuries very well. "**

took it easy on me. He's told me that he was very wary that if he pushed me too hard, too early, that I might just say, "This is bullshit. I'm walking." I was training every day except Sunday, week in, week out, for an extended period over the winter. Just getting my head around that was pretty startling, and I'd go walk about occasionally. I'd put in a few hard days of training, and then my body would be screaming, and I'd sort of think, "Oh, you can't expect me to just keep doing this all the time!" This coach was quite intimidating back then, so I'd go on these walkabouts and then be too frightened to call him. I wouldn't call him to say that I wasn't coming. I just wouldn't come. I'd leave it for a couple of days until I'd get the call at home, and he'd inevitably speak to Mum first, and then it would come down to me. Like I said, that first year, he just backed off. He didn't go through me. He sort of said, "Listen, I expect you to train, even if you're sore or whatever, you've just got to get used to this kind of regime." So, that first year, I thought I was training incredibly hard, but like I said, there were periods where I just couldn't hack it, but he was just one of those coaches who was really good at making me able to handle a lot more work than you could yourself. He's a genius. He's the best developmental coach in the world for the sport and still is; nobody is anywhere near him.

Listening to John talk about his responses to training—his body was "screaming," and he would choose to skip training without telling his coach—and listening to him talk about how his coach was intimidating, set off some alarm bells in my head. Watching and listening to John, at this point in the interview, I felt that he was trying to rationalize his own perceived shortcomings, that his coach's behavior and expectations were normal, and that John had problems because he could not always tolerate the training that was expected of him. He made it sound like his coach was patient and nice about everything, but the coach also appeared to have an agenda of getting John to do more work, to train through fatigue and soreness, and he could be quite intimidating. We have seen (in chapters 3 and 4) how complicated coach–athlete relationships can be. There is love, there is fear, and there is hate. Athletes idolize (and idealize) coaches ("He's a genius"), both the benevolent and the abusive. John feared his coach (thus the avoidance) and worshiped him too. For John, his coach was *perfect* even when pushing him past his limits, with the result that John engaged in passive-aggressive behaviors (e.g., disappearing). Often, the strategy of the abused is to passively aggress against the aggressor by not doing what the aggressor wants.

J: It still took me years to cope with the training rigors, and once I'd got that success as well, then coach started putting his foot down a bit. Once he felt like he'd hooked me, and he showed me I had potential, and I could get some good rewards out of this sport, then he'd start saying, "Well, listen, the Olympics are 2 years away. You know you have to train a lot harder than you've done there."

When John said that his coach had "hooked" him, he seemed to want to attribute his ambition to succeed at the Olympics to someone other than himself, despite his self-professed huge internal drive. It sounded like John's coach shared his Olympic dream, and was intent on showing John his way of achieving it, basically selling him on the hard-training path to Olympic success.

Having been a competitive athlete in several sports, I know well that success in sport often requires high volumes and intensities of training. Nonetheless, subtleties in the way the requirement to train hard is communicated to athletes can affect how they respond, positively or negatively, to training and how they make recovery decisions throughout their career. In John's case, it seems that he had a coach who introduced him to the concept of pushing hard but made him frightened to communicate about his fatigue. With respect to the coach's collusion in John's Olympic seduction, John hinted in the following comments that the Olympic dream *was* a shared one with his coach, and something that, originally, he and his coach kept to themselves.

J: We kept it a bit of secret, and he didn't tell anyone, but he had the feeling that I could make the Olympic team in 1992. I'd been gung ho; I thought I could make it too. Nothing was going to stop me; 1991 was the first winter that I really put in. I thought, "Yeah!" because like I said, Coach had mentioned to me, "Look, between you and me, I think you can make the Olympic squad." So, in 1991, I really put in as hard as I could and, yeah, it paid off in 1992, and I made the team.

It sounded like a lovely little secret, like one that might be shared between a father and a son. When John got sick at his first Olympics, his coach was supportive, like a protective parent looking after his sick, naïve son. In discussing the episode, John referred to the Olympic dream again as one that "we" (he and his coach) had shared.

J: Coach was trying to keep me up: "Don't worry. You'll be fine. You've done all the hard work." But I could see in his eyes that he knew, and I was still sort of young and a bit naïve, and I knew deep down that I was not going to achieve that long-shot dream we had.

Up to this point, John had given me the sense that his coach had been understanding, albeit demanding, with respect to training commitment and intensity. It did not take much longer, however, for me to see the origins of some of John's maladaptive responses to injury and fatigue, behaviors that seemed to have been learned from his coach.

J: Well, I probably didn't tell Coach about the injuries. I might have told a physio in passing that my knee was a bit sore. Yeah, that's one of the drawbacks with Coach; he couldn't deal with injuries very well. Well, he'd just get disappointed. He looked like a kid who'd dropped his boiled lollies [candy], you know. He'd get so down. He wasn't ever down at you. He wouldn't get angry at you, but you'd see him drop his bundle. He'd be just as upset as you were.

I took note here of John's description of his coach as one who would react to John's injuries like a disappointed kid. These comments about his coach seemed to contrast with those about his coach being intimidating. It made me think about how this dynamic might have affected John. I imagine it would have been confusing for John, on the one hand to adore and also be frightened of his coach, and on the other hand to see the coach act like an upset child. Like a child given the burden of a parent's anxieties, John was left to deal with his coach's childlike responses to injury and the guilt trip his coach was putting on him. I can see how it might have led John to keep quiet about small injuries and niggles. John went on to express his desire for a supportive coach, while acknowledging his coach's maladaptive attitude toward niggles and injuries.

J: You want someone to say, "Look man, it's all right. You're going well. We'll get around this." Although, that's not with niggles. Niggles he'd say, "Look, you live with niggles. That's what we do." But if I'd come to him and say, "Look, I've done something. I think my hammy feels tight," he'd be like, "Oh, shit!" Anyway, I probably gave him more grief with injuries than he's ever had with any other athlete, so it probably didn't help matters at all.

It seemed that the result of the coach's attitude towards John's injuries was that John blamed himself for giving his coach grief, not unlike a child blaming himself for his father's unhappiness. Moreover, John blamed his own physiology for being problematic with respect to injury susceptibility, rather than identifying maladaptive practices as causing the injuries.

J: I don't think my body type was great for triathlon, for the rigors of it. So that was one issue that I had with Coach. I gave him grief, but my knee was kaput in 1994. Again, I was just so shattered, I thought, "Now I'm finished." I did, for that moment. I sat up on the hill, and again I just cried, and I thought, "This is fucked. I hate this sport. I'm crippled at 24."

Here, John was again feeling shattered. The dream that was supposed to fill his void, his yearning for love and acceptance, turned out to be a bitter pill, an experience that left him shattered and crippled. The seduction of the Olympics promised so much to John, but brought with it so much pain.

Looking at his relationship with his coach, and how John had dealt with injury and fatigue throughout his career, it appeared that he had internalized his coach's attitudes. He had learned to react to physical setbacks, such as injury or illness, like a child who just wants to block them out, denying the consequences. His relationship with his coach had the qualities of taking care of, and being responsible for, a parent's happiness. Even in the best of parent–child relationships, children often take on the task of responsibility for parental happiness. This task is almost always doomed to failure, and in the case of a coach who is also a wounded child, the probability of success on this mission is close to zero. At the time of the interview, John was still training and competing, but had changed coaches (he could not meet the demands of fixing his coach's unhappiness). I queried him about his new coach's approach to injury.

Sean Richardson (SR): How has he been in terms of the injury stuff compared to your other coach?

J: Great. He sort of gets stuck into me as well. He calls me a pussy and all that sort of thing for breaking apart, but that's all tongue-in-cheek. Under all that banter he is really positive, and he will take you aside and say, "Look, you know you are okay. Don't worry about it." And, to me, it made a huge difference, so he's been really good with that. Like I said, he calls me a pussy and all the rest of it.

Here, John said that his new coach was great about injury, although he called John a "pussy," something that John mentioned twice, despite saying his coach was joking about it. Perhaps the coach liked to joke around with John, but given John's experiences with coaches and attitudes toward injury, I imagine that John might still have felt compelled, with this coach, to keep silent about pain, fatigue, or other niggles. "Pussy" here means "girl," something antithetical to "being a man." For many men, being a "man" is what sport is all about, and it appears that John had internalized maladaptive hypermasculine attitudes toward injury.

SR: How has this last year been, then, in training?

J: Awesome, best I've ever done. The best! I'm stronger and fitter than I've ever been.

SR: Any injuries?

J: Yeah, I just tore my calf 10 days ago. It turned out to be overload issues, like my calves were really tight from just doing a heap of running. They'd been tight all year, but they were particularly tight this week. While running I felt a bit of a pinch in my calf, or just underneath my calf, didn't really think much of it. I kept running, and it was hurting a little bit, but kept running. The next couple of days it was still hurting. Got to the point where I tried to run on last Saturday and felt something sort of pull a little bit. So, from then I've just been rehabilitating and strengthening, but that's really been the only major thing. I've had little niggles here and there.

In the previous exchange, I could not avoid seeing the contradictions in what John had said. He said he felt great—the fittest he had ever been—then a moment later admitted that he had just torn a calf muscle. I was affected strongly by what he said, I think, because I recognized attitudes and behaviors that I had seen many times in teammates and in myself throughout my athletic career. We are capable of deluding ourselves, thinking everything is great when it is not. It made me think that perhaps many athletes are programmed to give the impression that they feel good even when they are not necessarily healthy, a must-always-be-positive approach to life. Furthermore, John hinted that he still engaged in typical overtraining behaviors. Despite feeling pain and tightness in his calf, he continued to train until the niggles escalated into a more serious injury.

> **❝ One of the mantras in triathlon is, 'Rest is for the dead.' ❞**

At this point in the interview, I could see that John had been strongly influenced by his first coach and had internalized maladaptive attitudes toward training, recovery, injury, and illness. I guessed that such attitudes, and associated overtraining behaviors, were likely to be transferred to relationships with subsequent coaches. I had a picture of John—a motivated person, a self-proclaimed perfectionist—having early experiences with a hard-driving coach who did not like to talk about injuries. I had a sense, however, that there were other pressures, perhaps from other people, that reinforced John's relentless pursuit of Olympic glory. Talking about the sport of triathlon, John indicated that there are certain cultural expectations, and even stereotypical slogans, within the sport that support the more-is-better approach to training.

J: I mean particularly in a sport like triathlon where it's tough, I mean, you are talking about people who by nature have to be quite obsessive. It's a working person's sport. You have to do the work. It doesn't matter how talented you are. You know there is just no way around it, no other thing than to do the work, but you almost push to do the work without rest. You give up recovering. One of the mantras in triathlon is, "Rest is for the dead."

Given that he was surrounded by such cultural imperatives, I can imagine how it became difficult for John to make balanced decisions about training and recovery. For John, those sport-culture pressures were mixed with positive feedback from others and feelings of national pride, a potent combination of fear and inspiration. Reflecting on making his first Olympic team, John said that he loved the feeling of surprising everyone, and he was rewarded with recognition; people were impressed, and perhaps John was getting the love he so much desired. When he finally got to the start line at the Olympics, despite having been quite sick in the lead-up to the competition, John was totally fired up by feelings of pride for his country.

J: I knew I was flying before the Olympics, and when I got to the Games, I thought, "You don't need any motivation." When I walked out in front of the crowd and just to know I was in the Australian colors, I was so pumped up. That's probably why I was so shattered about everything, because I'd never been prouder. I could never have felt that proud, and I don't think I still ever have felt that proud since. Getting to the start, I was so aggressive. I remember because my training partner was there. We were lined up basically in a row, and we were like high-fiving, and we were like hugging and screaming at each other. "Come on! Let's just tear them apart! Let's just take them apart!" We were so pumped.

Is this scenario not what many athletes dream about, representing one's country at the Olympics? It had been John's dream. It had been my dream. As we found in conversations with experts, this seductive dream of glory, of representing a whole country's hopes, and of being the object of love and adoration, exerts a powerful influence on athletes. This desire to be the hero is also a source of stress that pushes athletes to exceed limits and engage in behaviors that ultimately may ensure that they do not become heroes. But who wouldn't be seduced? The irony here with respect to overtraining, however, was that John really was not ready to tear the world apart. He had been fatigued and seriously ill for the preceding week (the start of overtraining syndrome?) and was not physically ready to compete at the top level. Yet in the heat of the moment, he got carried away in his bubble of Olympic fantasies, believing, momentarily, that he was invincible—the delusions of grandeur creeping up again. Dangerous thoughts. With the contradictions between injury and invincibility staring me in the face, I asked John what he might have changed about those first Olympics, if he could do it again.

SR: How was your lead-up? Would you have changed anything looking back now?

J: No, I couldn't have. No way. It was just bad luck. There couldn't have been anything I could have done differently. I wasn't even thinking of holding back. I was going to go to the Olympics and then take another big step forward. It wasn't even, you know, bide your time, wait, wait, wait, get ready, and then peak. It was like I was still peaking all the way up.

John's answer to my question astounded me. Earlier in the interview, he had told me that his preparation had been great, but he had probably done too many hours at altitude that could have made him vulnerable to getting sick. A little later, he recalled that he had also experienced a relatively serious knee injury, in the form of tendonitis, prior to the Games. Nonetheless, responding now to my probing questions about his Olympic experience, John selectively blocked out those setbacks and claimed that he just had bad luck. The power of John's denial was overwhelming.

As the interview was coming to a close, John offered me one last peek into his fragile psyche, a peek that left me questioning the purpose of competitive sport, despite my loving it so much. Talking about a back injury before a World Championship, John pleaded with everyone to *fix* it, but he really came across as pleading for someone to fix him and all his shattered dreams.

J: That was the worst part of the lead-up to those World Championships because there was one spot in my back that I could point to all the time and say, "Look guys. There it is. Look, it's on that little joint there. Get it!" They'd try hammering that spot, then they'd try leaving it alone and hammering everything else around it. They tried taping it. They'd tape my ribs and back. Everything they did didn't work. That was the worst thing, because, like I said, my mind-set's always, "I'm competing. I am really ready." I knew I was in good shape, trained hard. I was going to do well at the World Champs, but the longer it drew on and the closer you got, the more I was thinking, "Come on. This is starting to get more serious. Okay, this is more serious now. Now, come on. Oh no! Shit! What's going on here? Look, they still can't fix it! What's going on? My training is suffering! It is just a niggle. It will be fixed. They'll fix me!" I made a decision to go to Germany to get treatment by the people I trust and know. "They will fix me." I had no doubts, but like I said, the closer it drew to getting there, the more the doubts were creeping in and the more realization I had that it's not fixing. Nothing they do is fixing it! I was really disappointed I didn't make the World Championships, but by then I'd exhausted every avenue. I'd given it absolutely everything, so I wasn't shattered, because the closer I drew to it the more I thought, "This has got to work this time! No it doesn't. No it doesn't. No it doesn't." It was like shooting a lame dog, you know; you had to put it down. So, to finish it, I went into the manager, and I said, "Look, you know you can't fix me. No one can fix me. That's it."

REFLECTIONS ON JOHN

"No one can fix me." Perhaps that was the final message about the dance with Olympic seduction. As we've noted multiple times in this book, some athletes look to this glorified event staged in front of the world as the solution to all their insecurities, as the ultimate achievement that will bring them happiness in life, the accomplishment that fills the void of existential angst, the cure for their internal wounds, or the one thing that will fix them. These athletes are misled, however, because sport is not there to fix them. The world of sport really does not care *who* wins, just that there is a winner. For John, as long as he used sport as a vehicle to compensate for his own feelings of inadequacy, he seemed destined to be disappointed. Ironically, even after all of these discussions and insights about the Olympics, as I looked over at John, I realized I would always love the Olympics. Despite everything that we had talked about, I still wanted to defend the glory of the Games, and maybe so did John. Perhaps

> **" It was like shooting a lame dog, you know; you had to put it down. So, to finish it, I went into the manager, and I said, "Look, you know you can't fix me. No one can fix me. That's it. "**

neither one of us was ready to give up all of our conflicting behaviors and attitudes, though at least we had become more aware of them at the end of our conversation. I also realized that the powerful seduction of Olympic greatness would probably continue to prompt athletes to surrender their better judgments about what is best for them and to overtrain. Nonetheless, I felt hope that by sharing stories of damaged bodies and crushed dreams, John might help other athletes find better balance in their lives. Maybe John's story would help athletes become more accepting of themselves and see that they do not need to be, nor can they be, fixed by competitive sport or Olympic glory.

How do we counteract the potential damaging influence of Olympic dreams that may lead athletes to self-destructive overtraining? An Olympic gold medal will never replace love denied or meet the need to be held emotionally. But we believe it will. We believe so strongly that a gold medal will heal us that we will thrash ourselves and damage ourselves to the point that Olympic gold becomes impossible. *Citius, Altius, Fortius* ("Faster, Higher, Stronger," the Olympic motto) sounds so wonderful, and if we can reach those ideals, then we will be healed. Love will be ours! Olympic gold, however, cannot fill the void. John's story reminds us of the "hungry ghosts" in Buddhism, a metaphor representing human craving and the behaviors we engage in to fill those cravings. The ghosts are always hungry because no matter what they eat, it is never enough. A person consumed with materialism may already have the beautiful house, the Italian sports car, the fancy yacht, the pedigree dog, and so on, but with each new acquisition he or she immediately starts looking for the next: "I will finally be happy (or respected, or loved, or fulfilled, or complete) once I have that summer house on the beach." In reality, that new house may bring momentary happiness, but it won't satisfy. Then it is on to the next material possession. The hunger never ceases. The hungry ghosts try to fill their bellies—swollen from malnutrition—but the effort fails to fill the void, bringing only more pain in the attempt. John seems like a hungry ghost who is trying to fill an emptiness, one that can never be filled, with a gold medal around his neck. Even if he had reached that goal, he would, we believe, still feel empty.

John's story disillusioned Sean, and he saw far too much of himself in John. He had been there. Though John is an aggregate case study, not an actual person, the stories and athletes that went into John helped Sean come to terms with his own tale of overtraining (see chapter 10). We hope John's tale will help others understand that seduction (with its elusive promises) is not love, nor can it ever fill the void many of us find when we look deeply into ourselves.

CHAPTER
9

THE PERFECT GIRL
Jane's Tale

When I (Sean) think about Jane, I reflect back on one of the perceptions that I held at the outset of this book about athletes who overtrain. I had the impression that there might be a specific type of athlete who would fit into a category called *overtrainer*. Steve and John both had elements of that persona (the supermotivated perfectionist who goes the extra mile, training through pain and competing while injured), but neither of them fit it entirely. Then came Jane. She was what I had imagined an overtraining athlete would be like. She appeared to believe, unequivocally, that *more is better*. She *hated* rest days. She responded to slumps, or any plateaus in performance, with increased training efforts. She went into complete denial about the consequences of injury, trying to block out awareness of parts of her own body. She did not want to listen to advice from others regarding recovery, and she constantly lived with the sense of not having done enough. Jane seemed to personify what it meant to overtrain; it was almost as if one might be able to isolate an overtraining gene in Jane's DNA. Nonetheless, I sensed that there had to be origins and explanations for Jane's traits and behaviors within her personal history. As she took me through her athletic career, which she was trying to rejuvenate after a crippling struggle with chronic fatigue, I learned that her overtraining (OT) had started at a young age and had been driven by abusive coaches and a pushy, overinvolved mother. Her story is one of an athlete exposed early to the pressures of competitive sport, a young athlete completely trusting her coaches to take care of her needs, an athlete who resorted to overtraining in attempts to keep her coaches and her mother happy, and to win their love and attention.

INTRODUCING JANE

As a child in competitive gymnastics, Jane absorbed the maladaptive attitudes and behaviors of her coaches to the point that she internalized them and they became part of her disposition, manifesting themselves in later sport and training environments. With a debilitating injury and a changing body in her early teens, Jane found herself no longer suited to gymnastics. Choosing to follow in her mother's footsteps, she took up the sport of track cycling, and her first cycling coach recklessly pressured Jane to thrash her body, teaching her that overtraining was the way to achieve success. Jane eventually left this coach, having been offered a scholarship to one of the Australian sport institutes. There, she had new coaches who seemingly took a more positive attitude, but who still reinforced her maladaptive training behaviors. Toward the end of her career, Jane managed to work with a coach who appeared to have a balanced approach to training and recovery. This last coach tried to dissuade Jane from her excessive behaviors, but by this time in her development she could not get away from herself, from her internalized slave driver. Like Steve's and John's stories, Jane's tale made me sad, but more than that it made me angry. Jane's story provided good justification for people to denigrate competitive sport, to point out that competitive sport can be the equivalent of child abuse. I had maintained my faith in the *goodness* of competitive sport, and here was a significant challenge to that faith. I took a deep breath and tried to relax, knowing that the next few hours with Jane would be challenging, possibly quite upsetting, but, in the end, fascinating.

Over the next couple of hours with Jane, I listened intently to the tale of how one develops an overtraining disposition. I listened to Jane's descriptions of abusive practices meted out by her coaches. I heard details of horrific injuries and of even more horrific responses to those injuries. I learned how Jane used dieting and weight control to win approval and to gain a sense of control over her body. I saw how overtraining, for Jane, became a vehicle of distraction from the loneliness and emptiness she felt in her life.

I AM AN OVERTRAINER!

At the start of the interview, Jane identified herself as someone who overtrains. She seemed to be proud of that identity, proud of being able to push herself when she was in excruciating pain or when she was overwhelmingly fatigued. I wondered if she was looking for approval from me for her overtraining identity, perhaps in a way similar to that in which she had looked to coaches for approval for pushing herself excessively. She came right out and announced that she is the kind of person who likes to thrash herself.

Jane (J): I don't believe in taking the easy road. I don't know the line. I'm a lot harder on myself, I think, than most athletes. You know there is the line where it hurts and you sort of go, "Okay, I'll stop now." I push through a lot more injury stuff and fatigue than probably most people would or should. If I do something, I want to make sure that I am the best that I could possibly be at that. It's not a competition against other people. It's purely a competition against myself, and I walk away thinking, "I could have done that better." If I pulled out and took the easy road, then I will cane myself in some other way. I mean, I am more motivated to go for a ride in 40-degree [104 °F] heat than in 30-degree [86 °F] heat. Explain that. That's totally weird. It's a mental battle with yourself to overcome everything. You just say, "I'm not going to let this stop me from doing what I want to do!" That's exactly

the same as when you're a bit sore. You just think, "That's okay. I'll just get over it." I push through more than most people, but that's probably habit from when I was younger.

Here, Jane seemed to revel in the identity of being a perfectionist who overtrains. Although she acknowledged that she probably pushed herself too hard, she placed herself above other athletes, suggesting that somehow she is tougher than they are, more motivated than they are. Jane's narcissism, her "specialness," would be a major theme throughout the interview. Narcissism is often the compensatory result of experiences of being degraded and put down early in life. People sometimes develop a sense of specialness to combat the hurt of being told they are not worthy of love. Thinking about Jane's identification with being a tough athlete, I wondered about what psychological or emotional needs she was looking to fill by being tougher than others. Jane did drop a big hint about the origins of some of her behaviors, beliefs, and attitudes when she commented that these were habits from when she was younger. I was left wondering about the experiences she had as a young athlete, and I was intrigued to hear more of her story.

Jane and I talked more about her tough attitudes and behaviors. Not only did Jane pride herself on doing extra training, she also admitted to loathing recovery.

J: You know, taking rest days, it's sort of not an issue; you just don't do it. I hate weekends. Weekends are the worst because everyone is meant to have Sundays off, and I am kind of like, "Damn it! Why do I feel this way?" I hate it. That's the biggest thing I've got to change, my attitude toward rest. I am hopeless. I never have recovery time. I just think, "If I just keep doing more, it's got to help," but I have got to stop that. While you are in it, it seems it's the right thing to do because you just want more, and more, of that thing that's going to make you better, but you don't understand.

Jane wanted more of that "thing"; I wondered what that coveted thing was about for her. I could not imagine that it was only about doing more training for training's sake, or even about getting better in her sport. In addition to pushing her training to excessive levels and avoiding adequate recovery, Jane admitted to having maladaptive responses to injury, illness, and performance slumps.

NO ONE CAN STOP ME

J: My response to injury was to work harder, train harder, which has the opposite effect. You are actually making things worse, but I didn't realize that at the time. All I knew how to do was, "I'm not going well. Work harder!" I always do that little extra bit, but I never told the coaches that I'd do it. I never said that I did this and that. When I had mononucleosis in year 7, I never told them that I was really tired. I'd just be grumpy, and in pain, and, eventually, got to the point where I just trained through my whole mononucleosis. Any performance slump for me was a sign to work harder. That was my response to a performance slump: Work harder.

Jane's working harder when injured or in a slump, and her hiding illnesses from her coaches, reminded me of John's tale in chapter 8. Training through pain and injury and hiding any weakness appear to be common behaviors in athletes who overtrain. In this quote, Jane's responses to injuries and performance slumps sounded defensive. She first stated that she knew she was making things worse, but then defended her training through illness with the mononucleosis example. She was going to train, and no one could stop her! Her responses also included elements of denial. She seemed to think that if she could just keep training, then she might be able to block out the consequences of injury. Along these lines, Jane described a time when she had been seriously injured in a cycling crash, but continued to train and compete.

J: My response was that, while my kidney was actually bleeding, while I was urinating blood, and everything, I backed off, because I didn't really have a choice. I was just too sick. The minute the contusion settled, against all outside advice and pressures, I just got straight back into it, and was still, obviously in retrospect, not ready to train, but I thought I was. I not only was training, but I was competing, and I think that was not good. But, you don't want to stop, and, as far as niggling type of injuries, I think that's part of the problem. I became so good at disassociation from the pain that I could never really get a handle on what was niggly and what was really disastrous. I

> **❝ I ignored it, and didn't look after it, and, come trials, I had problems lifting my legs. ❞**

ended up being able to block it out. As soon as I started thinking about it, unless it was actually like a broken leg and in a cast, then it wasn't significant. It just got wiped out.

As I was listening to Jane talk about her drive to do extra training, to push through pain and injury, and to avoid recovery, it was almost as if I could also hear a small voice of protestation, saying something like, "But I know better. I shouldn't keep doing these harmful things to myself." I had heard the same voice in my head when I was overdoing it in my sport, and my reluctant identification with Jane began to alarm me. There is nothing quite so unnerving as hearing parts of one's own pathological history echoed in the words of another. At the time of our interview, Jane seemed to have reached a point in her development where she had been educated enough in sport science to *know better*. She had made enough mistakes from which to learn, and she had a pretty good idea of what was right for her body in terms of training and recovery. Nonetheless, she was constantly battling herself, fighting some internal struggle to tame a pathogenic drive toward excessive training and self-harm. It seemed that her way of coping with that internal struggle was to cut herself off from the elements causing the conflict, to deny or downplay the consequences of her excessive training, and to block out the realities of her injuries and other setbacks.

Jane's maladaptive attitudes toward injury manifested in many ways. In the following example, Jane illustrated how she had downplayed the severity of injuries and attempted to mask the pain with drugs so that she could continue to do her sport.

J: I landed directly on my sacrum, and the track was really hard, and I just smashed my sacrum, and prolapsed two discs. It was a compaction injury. It just went bang! My response to that was unbelievably glib. I couldn't sit down, and I couldn't lie down. I could kneel. However, I decided that I had a periosteal bruise because I've had periosteal bruises before, and they're as painful as fractures, and so that's what I decided I had. I went down to the local hospital, and there wasn't a doctor there, and they said, "Do you want us to call one?" I said, "No, just give me something because I've got to get to a race." So, they gave me painkillers, and I just dosed myself out. I knew I was a mess, but I knew that if I'd sat down and thought about it, logically, I wouldn't have been able to do it. I just so badly wanted to keep going that I just tried to say, "No, no, it'll be fine; it'll be fine." Because it's, you know, your dream. So, with the injury, I just tried to ignore it. I ignored it, and didn't look after it, and, come trials, I had problems lifting my legs.

In this description of a crippling injury, Jane talked about making her own diagnosis, one that minimized, at least in her mind, the potential damage of the injury, and allowed her to continue training and competing. Jane also admitted to ignoring the advice of her doctors, believing that they were likely to tell her to take more rest than she believed she needed.

J: I used to think about medicos: If you did what they said, you'd never get out of bed. You know what I mean? I used to have that feeling that they always tell you to take more time off than you needed. Everybody is always overreacting. That might've been me. That might have just been my attitude towards external control, and I had high personal control needs. So, when you've got somebody telling me to go to bed, they might as well tell me to shoot myself in the head; that is where I will just atrophy. . . . It's a horrible thought.

Jane created quite a powerful picture here, equating bed rest with suicide. If she could not train, she might as well not be alive. She would "atrophy" and fade away to nothingness. Training was her identity, her *self*, and if she didn't have that, then she was nothing, an empty vessel. That fear of loss of self had to be constantly combated with more bouts of training so that Jane could feel she existed and had purpose and meaning. Once again, I was left wondering about the origins of such strong emotions. At this point, Jane offered a piece of the puzzle. She gave me an inkling that some of her maladaptive responses to injuries, illnesses, and fatigue were part of coping mechanisms she had learned in her family unit as a child in order to deal with a debilitating illness and parental distress. She also learned many other early lessons from her parents about relationships and love.

PARENTAL INFLUENCES

At this point, I could not help thinking about what Jane's relationship must have been like with her parents. I wondered how her relationships with her parents might have affected the dynamics she developed with coaches throughout her career. Jane's descriptions of her parents, especially her mother, left little doubt in my mind about what I had begun to suspect. It had seemed to me that Jane's attempts to please her coaches (see the Early Sport and Performance Lessons Learned section later in this chapter) had been a transference of a dynamic she had learned at home, possibly one in which she felt her parents' love was contingent upon her success in sport. Jane started off by telling me that she felt her parents were overinvolved in her sports.

Sean Richardson (SR): How were your relationships with your parents through your sporting career?

J: Well, I thought it was always pretty good, but I always thought they were too involved, like Mum being on the administration of the cycling association and Dad the president of the club. I didn't want them to be involved, but that's what happened—just too much. I like to have my family, my social life, and my sport all different. I don't like to combine them. It just annoyed me, the fact that my family would be involved that way. I just wanted them to be supportive of me, just me, not everyone else.

Jane had mentioned earlier in the interview that she used her sport to distract her from the rest of her life, to fill the void left by her loneliness. Talking about her parents, Jane seemed to have been crying out for love, crying out for her parents to save her from her loneliness, and to focus attention on her, not on her sport or "everyone else" in her sport. These cries—"be supportive of me, just me"—sounded more like *love me for me, just me, not for what I do.* I think this need is at the core of Jane's maladaptive thinking and behavior. Underneath it all, she did not want to be loved for her sport performance, but for herself, and she resented that sport was her only way to gain attention. She wanted to be held and loved for being Jane, but when it seemed that love and attention were contingent on performance then what choice did she have? Jane learned that just being Jane was not good enough, not worthy enough, and she sought to fill her needs in the only way that had worked in the past: She trained and trained, then overtrained, ensuring that the thing she feared most would happen. This maladaptive and self-defeating pattern—of behaving in ways that make sure one does not get what one needs—is a major theme I have observed in working with athletes and performers.

In her next comments, Jane told me about the importance her mother placed on her cycling, and how her mother made her feel unworthy of attention when she was not training or competing.

J: I thought my parents were really good until I realized Mum was sort of more pushy. When I quit cycling, she wouldn't talk to me, and then, as soon as I went back, she was my best friend, and took me shopping. That sort of hurt me. It's the silence that makes you feel like that. It was sort of just like, "Right. So this is, obviously, what she wants me to do." I did not necessarily feel accepted, but that is how she'd be happy with me. Now that I haven't raced for two years, it's been really hard. We are still not getting along. She's a nightmare, an absolute nightmare.

It was heart-wrenching to hear Jane talk about her mother. I could see the pain in her eyes. She appeared to love her mother, but she did not understand how she could make her mother happy. She could not find a way to feel accepted or loved. Jane's response to her mother's unreliable and contingent love had been to turn to her coaches for love and affection. She would do anything, including train excessively and push through horrendous injuries, to keep them happy. Jane's mother had also modeled overtraining behavior in her own athletic career, having pushed through debilitating injuries on several occasions.

J: I think I come from a family that copes very well. It is very solution orientated. I have a mother who broke her neck cycling, and had three fusions, and her coccyx removed, and, yeah, she never says die. I think that's been a role model.

Jane's perception that her family copes well seems seriously skewed. Her mother's injuries stopped a career, and now she was trying to live out her own dashed dreams through her child. Her love was contingent on Jane's performing well, and there appeared to be a lot of mum

in Jane. She internalized her mother's values, and those values became her default mode. I could see now how well some of her coaches had fit, hand-in-glove, into her model of a relationship. What Jane learned from her mother was that the response to injuries was to never say die, or, in other words, to ignore them if you possibly can. Furthermore, at times, her mother even questioned whether the injuries were real or were created in Jane's mind. Jane described how her mother seemed to believe that Jane used injury as an excuse to discontinue her sport.

J: I think, at the start of my back problems, they didn't realize the severity of it, and they just said, "Oh, is it psychological, or is it really an injury?" Because there wasn't a real answer to what the problem was, originally. I sort of felt that she thought that I was just making excuses not to cycle anymore. I know she really wanted me to go overseas, and continue my scholarship, and everything like that, and I really wanted to go as well, but it just came to the fact that I couldn't race or train, and I needed to compete to be there. Yeah, so, Dad was always good about injuries and stuff, but Mum was more on the pushy side. Friends and people had always asked, "Oh, does your Mum push you?" I thought, "No, no, no, she's being supportive," but, when I stepped away from it, I could see it. She still, now, wants me to be a little world champion.

> **" I thought my parents were really good until I realized Mum was sort of more pushy. When I quit cycling, she wouldn't talk to me, and then, as soon as I went back, she was my best friend... "**

Even though Jane said that she wanted to continue cycling, and to go overseas to train and compete, it sounded like her underlying motivation was not about pursuing cycling, but about pleasing her mother. Furthermore, even if she could please her mother by being a world champion, she would still be a "little" world champion, as if she were saying, "I can never be grand enough for mum, no matter how hard I try to please her." Sadly for Jane, instead of feeling that she had pleased her mother, she felt accused of faking an injury.

I wondered how a mother could be so hard on her daughter. Perhaps it was because mum was thinking about herself, not about her daughter. Jane had told me that her mother had a big disappointment as an athlete; she had ended her own athletic career early because of a serious injury, which shattered her dreams of being a world champion in cycling. From Jane's descriptions of her mother's pushy behavior, it sounded like mum was driving Jane to heal the wounds of her own past.

Trying to fix her mother was a powerful motivator for Jane. I could only imagine the pressure resulting from the combination of Jane's wanting to fix her mother, and mum's overtly pushing her to do it. This pressure left Jane full of confusion about her own motivation to participate in sport.

J: I think mum's motivation comes from the fact that she knows I could be the best in the world, like she had hoped to be, and she just wants me to fulfill that, because she knows that's what I want to do, but I don't believe in myself, because of my injuries and stuff. So, I think she just wants the best for me, even though I don't think she does sometimes, but she's a parent.

Jane was full of contradictions here. She sounded like she was trying to make excuses for mum, but under it all she sounded angry with her mother. She seemed to suggest that it was okay for mum to push her against her will. It was okay for mum to make her love contingent upon Jane's success in sport, as long as the motivation was to prove to the world that Jane was the best. Jane's mother had dreamed of being a world champion, and maybe Jane's desires for respect and attention from the world have roots in trying to fix her mother's disappointments: "If I fix mum, then maybe she will love me for me. When the world loves me, then mum will *have* to love me too." It was all okay because mum wanted the best for Jane, but in reality she wanted only the best for herself, and it was not okay for Jane. Looking at Jane during this part of the interview, I could see nothing but pain and confusion.

Jane's mother had also become so involved in Jane's sport that she went out and taught herself about the physiology and the mechanics of cycling over and above what she had learned as an athlete. Despite not being a coach, or part of any support staff, she tried to explain to Jane how and why Jane's body would respond in different ways, seemingly trying to justify pushing Jane to train harder.

> **" I just wanted her to be a Mum, to be there if I crashed, but she wasn't. "**

J: Mum would try to explain what was going on in my body, that it was an adaptation to get over, but you don't listen to them. You can't understand, being so young. You can't understand what's happening. So I was like, "Yeah, whatever Mum." But I just wanted her to be a mum. I didn't want her to be an expert. I just wanted her to be a mum, to be there if I crashed, but she wasn't.

Jane had just wanted a mother. It sounded simple. She wanted love, but her mother gave it to her only if she competed and was successful in sport. Thus, in training, it seemed natural for Jane to seek out love and approval from her coaches as potential substitutes for her mother.

Up to this point, Jane had focused on her mother's pushing her to compete, and I was left wondering how her father fit into the picture. She had mentioned earlier that she felt he was overinvolved, being the president of the cycling club, but she had said that he was "always good" about not pushing her. When I asked Jane again to reflect on the origins of her drive to push herself, she began talking about her father.

SR: Where do you think your desire to push yourself came from?

J: That's just me.

SR: That's just you?

J: Yep. I think my dad is fairly driven. My dad's the same. He pushes himself, and he is a hard worker, and totally motivated. Dad is like me, but not to the same degree. He's probably another step down. I'm pushing my limits, but he's so much like me, and that's where it comes from. I think I'm a little bit of the son my father never had.

Jane identified with her dad. Maybe she thought that by emulating him, or being even tougher than he had been, she would impress him enough to make him love her more. Perhaps Jane thought that if she could be more like a boy, a tough boy, ready to push herself like Dad, then he would love her more. Jane did not tell me more about her father, but I had heard enough from her brief comments to see that some of her overtraining behaviors were also manifestations of a desire to please him. It seems Jane learned that she would get her dad's attention and love if she were more like a boy, became tough, and hid her weaknesses.

HIDING WEAKNESSES AND BATTLING ANXIETIES

J: I had a serious immune deficient problem when I was young, and my quality of life was very good because I never said die, and because I kept my spirits up and everything. I didn't have the consequences, socially, of chronic illness because I would die in private, and I'd be up around others. I could switch off, and I think I became very good at that.

In this example of coping with an immunodeficiency problem, it seemed that Jane believed she could will herself to get better. Perhaps, the fear of being debilitated by illness was so overwhelming that Jane's only way to cope was by cutting herself off from it, or at least by pretending to others that she was okay. Perhaps her parents also reinforced Jane's silence and blocking out of the pain by rewarding her for being a positive, happy girl. Perhaps Jane learned that being tough and independent was the way to win more love from Mum and Dad (see the previous Parental Influences section in this chapter for her desire to be loved). In her next comments, Jane repeated her dislike of being dependent on others.

J: When it came to my injuries, with the kidney and the ribs, I very much didn't want to feel any dependency on anybody. I just wanted to be on my own, do my stuff, get on with life, and, of course, being stuck in bed is not conducive to doing that. You get to lie there and think about everything. I think I was trying to avoid the situations where I would be lying in bed doing nothing, relying on other people, and thinking about everything.

This quote is the third time Jane has talked about being afraid of being stuck in bed. It seemed like her early experiences with being ill and incapacitated had been so frightening to her that she would do anything to avoid future experiences that resembled her childhood immunodeficiency illness. It seemed that Jane's antidote for this fear of being incapacitated was to do anything to maintain her identity as a healthy, hardworking

athlete, including training through illness, injury, and debilitating fatigue. Eventually, Jane used training as a way to escape, not only from fears of being sick or injured, but also from fears that her life was empty and meaningless.

J: I came here just for cycling. I put so much heat on throwing myself into cycling to distract myself from the rest of my life, and going, "Oh well, I'm not that happy. It's a bit lonely." You know, I don't have many other things; so I was overtraining myself, you know, with my coach having to say, "Back off." I think I just wanted to do what I do. I can't put it more simply than that. I wanted my life just to keep going, and I'm good at what I do, and I just wanted to do it. So, I was just completely ignoring, not only everything everybody said to me, but I was totally ignoring my body. That's my nature, in terms of training, to distract myself and take my mind off things. So, I fell into that trap myself, not from any pressure outside, other than me saying, "Oh well, at least it will take my mind off how bad other stuff might be." That's probably going to be the hardest thing for me to change, just getting back to training because I love it, and not because it's a distraction from my life.

I must have been wide-eyed as I listened to Jane talking about using training as a distraction from the rest of her life. It sounded awful, painful, and depressing. I have often heard myself say to others that training is a great way to meditate, to process the stressors in my life, and to let go of them, but I do not think I have used training to forget about my life. Earlier, I had speculated that Jane used sport to fill a void. From this segment of the interview, it seemed that Jane had been grasping at her sport as if it was the *only* thing that could make her feel okay. Unfortunately, this maladaptive attachment to sport was manifested in all kinds of damaging behaviors and activities.

THE GRAVITY OF WEIGHT

Jane's desperate desire to control her life and win approval manifested in extreme dieting. Caloric restriction was a behavior for which, in her last year as a gymnast, she had been rewarded with praise and recognition. Then, at age 15, she got caught up in thinking she would be faster on the bike if she could increase her power-to-weight ratio by dropping weight.

J: I'd never been that disciplined before, and I was really, really pleased with my effort, and, you know, when I lost 8 kg, everyone noticed, and I think that really made me keen. People really recognized your work. Obviously, it can't just fall off; you have to really do some work to do that, and it is really another way that people can recognize how hard you worked. It was extreme dieting, but it worked beautifully! I felt like I was asserting myself, athletically. When all the other athletes sort of dwindled off, and didn't do anything for the rest of their seasons, well I'd really redirected my efforts, and, you know, done something good about it. I think that's important when you are working that hard. People say, "Hey!" You want people to notice that you are doing good work.

Jane had turned to weight loss as a gymnast to win approval, to get recognized, and to be loved. Now, as a cyclist, she said that the extreme dieting had worked *beautifully*, and people noticed her. Three times she commented that weight loss was a way for people to recognize her hard work. She seemed driven to gain this recognition, and I wondered what happened to Jane when people did not recognize her hard work.

In the case of her dieting behaviors, the praise and reward for the weight loss were temporary. The behavior became problematic when Jane tried desperately to repeat the perceived positive outcomes generated from her first experience with weight loss as a cyclist. With her changing adolescent body, and her continued work as a track cyclist, the internal and external pressures to diet and keep skinfolds low began to negatively affect her fatigue levels and performance.

J: In subsequent years, when I'd try to diet and lose weight, my performance would really drop. I was never totally successful in losing that kind of weight again because every time I lost weight, or was really focusing on dropping weight, my perfor-

mance would just really drop. It got to a point where I was so tired that I started to question everything about cycling. I guess, with the eating thing, it was to get approval when I was at the institute. I suppose it was, continually, to show that I was good enough.

Jane was trying to engage in the same behaviors that had worked previously to win approval from people around her. Now, however, her performance suffered when she tried to control her weight. Coaches would only love her if she was performing well, but her cycling coaches at the institute did not see the weight loss as undesirable; rather, despite Jane's decreasing performance, they pushed her to bring her weight down even further.

J: The coaches were constantly saying, "Lose weight, lose weight," even though I was the smallest one there. They just wanted me to continually lose weight, so I stopped eating altogether, and lost a lot of weight. No one, not one person, even took into consideration why I had lost so much weight. They didn't even say, "Hang on, what's going on?" I thought that was really strange, because, when I came home, all my friends, and family, and athletes that I train with here, they just go, "Oh my God!" whereas the institute staff didn't even come up to me and confront me.

I could imagine how debilitating Jane's disordered eating must have been. She participated in a sport that demands high levels of energy output. Cycling *requires* athletes to ingest high-calorie diets. Without eating enough food, Jane would have struggled enormously with sustained training efforts. It seems incredible to me that the coaches kept pushing her to lose weight in the face of her decreasing performance. The result of Jane's weight control behaviors was that she could not perform at all or continue to train at any proficient level.

> **" The coaches were constantly saying, "Lose weight, lose weight," even though I was the smallest one there. They just wanted me to continually lose weight, so I stopped eating altogether, and lost a lot of weight. No one, not one person, even took into consideration why I had lost so much weight. "**

J: When I was dieting, I couldn't concentrate, and I didn't have enough energy to train well, and so it didn't really matter how light I was, you know; it wouldn't make any difference because I wouldn't be able to train well. Well, probably for about 6 months, I was struggling more than usual, but you know I was dieting. I was used to being tired, and I was training harder.

Jane had been allowed to continue on her destructive path within an environment supposedly designed to support athletes and provide them with guidance and knowledge about health, training, and recovery. How could the coaches not link her poor performances to her dieting behaviors and weight loss? Perhaps Jane had kept a lot of information from her coaches, but I was still left wondering why they did not investigate the reasons behind her declining performance.

At this point in the interview, I thought we were ready to delve further into Jane's story, to examine her relationships with coaches and parents, and to get her reflections on being a competitive athlete for most of her life. Having been introduced to Jane's pathogenic behavioral tendencies with respect to training, recovery, and nutrition, and having been given a few hints about possible origins of her behaviors, I was eager to gain more insight into what had brought Jane to her current state.

EARLY SPORT AND PERFORMANCE LESSONS LEARNED

As with Steve and John, Jane had enjoyed tremendous success at a young age. She had been swept into the system of sport institutes in Australia, where she was showered with praise, promise, scholarships, expectations, and pressures.

J: I was 4, doing gym until when I was 12, and I was State Champion in gym and bars. I competed for Australia from 8 till 10, winning Pan Pacific Games, and then Junior Worlds, but at age 12 I broke my arm, so was forced out of gym due to injury and started track cycling. I followed in Mum's footsteps when I was 12. Cycling took off and, within 3 weeks, I was State Champion. Three months later I was Australian Junior Champion. I was Australian Junior Champ for years 12 to 14. I was asked to go to the institute when I was 13.

Then after a few down years I was Australian Under 17 Champion for Olympic sprints and pursuit. I won an international event at 17. I made the youngest ever, my only claim to fame, Australian Open track cycling team at age 17 or 18 or something, and I was training morning and night. Then, with missing Atlanta by 3 seconds, everyone realized that maybe I was good enough to go to Sydney.

That Jane considered winning an international event at age 17 her "only claim to fame" seemed odd in light of her other accomplishments. This comment, however, is congruent with Jane's approach to sport. Almost any accomplishment is never enough; nothing can really fill the void except for the transient relief of training.

Reflecting on her time at one of the institutes of sport, Jane recalled feeling a lot of pressure and expectations, especially given her meteoric rise as a junior.

J: I had always gone to the institute with such high expectations, after being ranked so well as a junior, and I was just, automatically, going to make this magic transition from junior to senior, and take over the top spot. When it didn't happen right away, all of a sudden they were going, "Oh well, maybe she's a dud." They were kind of rethinking, changing their minds. I mean, going to the institute was good, but there was a lot more pressure on scholarships, which I hadn't had before. So, constantly, if I wasn't achieving certain performance levels, I had the heat on me. There was also a lot more pressure on weight, skin folds, all that kind of stuff, which wasn't enjoyable.

Going to the institute was good, then it wasn't. Coaches were fantastic, then they weren't. These confusing and conflicting perceptions and experiences pervade Jane's story. Something (or someone) happens that temporarily fills her void and helps her feel that she exists and is loved, but then that something sours because it is not the medicine that can cure the emptiness. Positive events and caring others give only transient symptomatic relief.

Jane wanted *magic* to happen, and there seemed to be no room for failing, or even faltering. When Jane did not live up to the expectations of her scholarship coaches at the institute, she believed that they no longer had faith in her capabilities. Given Jane's harsh internal critic, I was not sure whether

her perceptions of her coaches were reflective of reality, whether they really did treat her poorly when she did not perform well. Nevertheless, Jane seemed convinced that being loved and accepted by her coaches was contingent on her level of success in her sport.

J: At the start, I came and the coaches loved me, and I was the best, you know, the next hot thing, and then as performance dropped, all of a sudden, I wasn't so great, and they weren't so interested anymore. It's a real bandwagon thing at the institute; they jump on and off like anything. I was out of favor when I hadn't come on within a year. When I hadn't automatically made World Champs, or Commonwealths, or performed, they didn't want a bar [anything to do with me] of me. So, I'm pretty sensitive to that; I am aware of what's going on.

Jane was sensitive to being out of favor, to suffering neglect from her coaches, and to feeling uncared for and unloved. Given Jane's massive attachment to her athletic identity, it made sense that she was so sensitive. Nonetheless, I had to ask myself again: Was Jane overreacting to her coaches, or were they really that harsh with her? On the one hand, having consulted with athletes in a number of sports, within a sport psychology context, I have seen athletes who are hypersensitive to coaches' behaviors. Such athletes are likely to take almost any critical comments made by a coach as personal attacks, even when coaches do not intend them to be interpreted that way. On the other hand, I have also seen several coaches dish out harsh, derogatory, and isolating criticism to athletes. In the end, it does not matter what the coaches actually did or said. We are looking at Jane's world.

Jane talked about her first, longtime gymnastics coach (we'll call him Matt) with fondness. She did not identify his style as pushing her to overtrain. Rather, she recalled that Matt babied her, perhaps taking too much care of her in such a way that she did not develop knowledge about what was best for her body. Jane also did not learn to think for herself.

J: Oh, he was awesome. I got attention all the time, and did not have to think very independently—not having to self-analyze

> **" Even though Matt loves me to bits, he still didn't want to hear about the injuries. "**

what was going on, you know, what's the best thing to do. I was babied, which was great at the time, because I was looked after really well. This is why I think Matt was so good. He never really asked me if I was tired or hurting, but he could tell when I was irritated or a bit pissed off.

Jane appeared to have idolized (and possibly, idealized) Matt. She received a lot of attention from him, and, I imagined, a lot of care and love. Nonetheless, I felt that despite all of her praise for Matt, Jane might have a few negative things to say about him as well. That Matt did not ask Jane about being tired was my first clue that things with him were not always "awesome." Perhaps he did not want to know about any of Jane's physical complaints. In the following comment, Jane said that she and Matt were the perfect team. It sounded like she might even have perceived their relationship, on an unconscious level, as that of a perfect, loving couple, or perhaps of a perfect father–daughter union. Her relationship with Matt deteriorated, however, not long before the end of Jane's gymnastics career.

J: Matt and I were so, so close between ages 8 to 11. He was so good to me, and we just got on perfect, a perfect team, the two of us. Then to this day, I still don't know what happened. When I was 12, we just started fighting and arguing, and I'd get kicked out of training, and he just wasn't coaching me how he used to. I think he had personal problems; maybe something happened in the family. He never talked about anything. It really hurt me. So, um, yeah, and I just wasn't getting any training done because I just kept getting kicked out of the gym. We just didn't get on at the time. It's like how parents don't get on with kids.

These mixed-up feelings of love and disappointment, perfection and flaws, seem to be repeating themes that started for Jane at an early age. Hearing about the demise of Jane's relationship with Matt, I got the sense that she had been putting Matt on a pedestal, fantasizing about how great he had been to her, when in reality it might have been different. Jane revealed that Matt was not very communicative; instead of working things out with Jane when tensions arose, Matt chose to kick her out of training. I took note when Jane suggested her interaction with Matt was similar to a parent–child interaction. It seemed that she was looking to her coach to fill the role of a loving parent, perhaps a role neglected by one or both of her own parents, and when the coach deviated from Jane's idealized image of the perfect parent, she was devastated.

When I asked Jane to expand further on what happened with Matt, she gave some information suggesting that he was not always so balanced in his approach, not always *perfect*. Like many of her coaches to come, Matt did not want to deal with Jane's injuries.

> **J:** When it comes to injury, he just didn't want to hear about injury. Well, all my coaches have said it, my institute coaches, and even Matt. Even though Matt loves me to bits, he still didn't want to hear about the injuries.

I thought about how Jane must have been in conflict in her relationship with Matt. On the one hand, she talks about the relationship being perfect, portraying Matt as the perfect father figure who loves her "to bits." On the other hand, he does not want to hear about her injuries, and he communicates poorly with her when problems arise. Jane loved Matt, but she also said she was extremely disappointed because he did not take care of her and withdrew when she was injured. These confusing emotional reactions in interpersonal dynamics are what happen when love and attention are perceived as contingent. In Jane's world, Matt was wonderful when she was wonderful (performing well), and Matt was not there when she was injured or in a slump. I imagine that Jane might have developed an insecure attachment to Matt, one characterized by confusing messages and contingencies. Jane was rewarded with love and praise for being the *perfect* athlete, but shunned for being a little bit fallible, for being human. It does not seem surprising that Jane resorted to overtraining behaviors in attempts to please her coach and keep receiving the love she dearly craved. Jane ended up leaving the sport of gymnastics, having learned to rely substantially on her coach for guidance and support, yet also receiving the message that she should keep quiet about injuries if she wanted to receive the coach's love and attention.

Jane was confronted with a whole new set of demands when she moved from gymnastics to track cycling. She had been bitterly disappointed with the loss of her intimate relationship with Matt. Perhaps she was seeking to fill a new void left by Matt's absence. Maybe she was seeking out a new father figure, a new object of attachment, someone she could please, and from whom she might receive love and attention. Unfortunately, Jane's next coach reinforced her previous overtraining behaviors and seemed to instill an even more intense commitment to training excessively and pushing through injury.

> **J:** Yeah, basically I was training way too hard. My first cycling coach, Jack, wasn't very experienced, and he was just thrashing me. He was a bit of a psycho. A lot of it was overuse kind of stress and injury stuff. Jack was very theatrical and very, very abusive, actually, which I was pretty used to because my sister was a dancer, and ballet teachers are cruel. But, at the time, I saw that as quite a good thing because, generally, if they're really on your back, and they're pumping you hard, it's because they think you'll make it. My coach's attitude was, "If you can't cop this, then you can't cop the pressure of the Olympics." So, although it would get to me at times, it actually worked well for me because it would make me angry, and that anger actually was quite a good energy. So, yeah, I felt strong by being able to cope with it, whereas other people were in tears, wanting to give up. I felt quite virtuous that I could cop the abuse. Besides, I was getting results, and I thought, "Oh well, maybe this is what it's all about." Also, I didn't know any better. I just thought, "Oh well, this is how it is with elite sport."

> **" *My first cycling coach, Jack...was very theatrical and very, very abusive...* "**

We have seen this Darwinian, survival-of-the-fittest approach to developing and coaching championship-quality athletes over and over again. Thrash those athletes in your charge, and the ones that remain standing are your champions. Trisha Leahy spoke about this abusive weeding-out process in chapter 6, and Steve's tale in chapter 7 showed how cruel Darwinian professional sport can be.

Jane was still so young. She was new to the sport of cycling, and she was accustomed to having a lot of attention from her previous coach. In gymnastics, Matt had supposedly looked after her, but he did not help her develop any sense of autonomy or any knowledge of self-care. Then, in her new sport, she had run head-on into a coach who would give her attention, but only if she learned to cope with his abuse, put her head down, and trained hard.

Comparing Jane's first two coaches, it seemed like the dynamic with Matt was based on mixed messages: praise, attention, and affection when she was healthy and training well (what Jane craved), and neglect and perceived unworthiness when she was injured (what Jane feared). Matt was more passively abusive with his neglect, whereas Jack was consistently and overtly abusive, filling her training with fear and insecurity, and Jane coped by internalizing overtraining behaviors and an I-am-tougher-than-everyone-else mentality. It seemed to make sense to her because, as she commented, she achieved good results with Jack. Thus, Jane created and tried to live up to the identity of an athlete who could endure more than any other athlete (more narcissistic specialness). Nonetheless, Jane acknowledged that she did not enjoy her training very much at that time of her life.

J: I don't remember really enjoying it that much. I was permanently tired. The injury stuff was quite freaky, and there was always a lot of pressure. If I strained a hammy, it was like, "Oh my God!" I ended up having to hide a lot of injuries and stuff because you know he just couldn't cope with it. So, stuff like rushing around to acupuncturists getting treatment, trying everything to get better, and I ended up training through a lot more injuries than I ever would now. [Jane's mother played a large role in her overtraining and injury treatment behaviors (more about this later in this chapter).] So, it was kind of a blessing, actually, at World Juniors, when I crashed in my race, and he was so mortified. He didn't even come up after me and see if I was all right. He just left the country, went off to France. And I came back home after that, and I can't remember how long I had off, but the injury definitely got worse. It may be something important to consider, too, the psychological and emotional issues that come with being injured. Okay, I have got an injury, but, when I got back, the pressure and being forced to train, and being forced to race, what that does to you emotionally and psychologically, and how that might affect your body, and make the injury worse.

Anyway, it was horrific. Frustration, the anger, the depression, and everything like that that comes with injury, plus on top of that having a coach demand that you train. Now, there's a trust issue that maybe he doesn't trust you at times. He didn't trust the physios. He thought that the physios were putting extra weeks on and trying to get me out of training. He had this impression that at times I was just a bit lazy and wanted to get out of training. Yeah, he was a nutbag! So, heaps of pressure, and he didn't know how to cope with failure or injury at all. I was constantly pretending I wasn't hurt.

Jane talked about it being a blessing that her coach left her alone after she crashed in a race in Europe, but I wondered if she might also have had feelings of being abandoned by Jack. It seemed that Jane was given the message that if she wanted the coach's attention, and if she wanted his trust and his love, she should keep quiet about her pains and injuries. She called Jack a "nutbag," but I sensed that she was still quite attached to him. That attachment seemed to take the form of solidly internalizing Jack's values and (self-)abusive training practices, a type of disordered attachment and identification with the aggressor. For more detailed examination of this process, see chapter 6.

Eventually, when Jane realized that she would have to leave Jack, she did admit to having a difficult time doing so. Despite the pressures, coercion, and abuse she had faced with Jack, she suffered when she had to separate herself from him, perhaps not unlike a victim finding it difficult to leave an abusive intimate relationship.

J: That was probably the biggest thing I had to do. I knew that, and I was miserable. I had to leave him, and that was awful, because he got me, basically, from nothing to being number one in the world. I knew he was doing the wrong thing by me in terms of overtraining, and I was just exhausted, you know, 13, 14 years old, and a mess. So, leaving him was probably one of the biggest things I had to do in terms of changing. He went crazy when I left him. At one point, he just got so angry that he just took a chair and threw it at me. Jack was just a nutbag.

Jane acknowledged that Jack was pushing her too hard, and that he was abusive and aggressive in his behaviors, but she still believed in his method because she had been successful under his coaching. She had still been drawn to him, had still looked to him for love, and was crushed by the thought of leaving another parental figure (an attachment probably based on her relationship with her mum, see below), even such an abusive one. After training in this environment permeated with fear, tension, and pain, Jane internalized much of Jack's slave-driving mentality. This internalization of abusive coaching practices then became all too complete; Jane moved into her next two training environments as the abuser. The object of her abuse was herself, and the method of abuse was overtraining.

INSTITUTIONAL ABUSE

Jane's next move was to one of the institutes of sport in Australia, where she had been awarded a scholarship to train full time. Jane was excited at the prospect of having a system of support and the country's top coaches to guide her.

J: So, it was a new start, and it was pretty motivational because you know you're there with the best in Australia, and you've got the best facilities, all paid for, and everything was great, but still. Yeah, I went up for, I think, a couple of months, like a trial. The coaches were so nice, and everything was perfect, and so I was like, "Yeah, yeah. I want to go back up." So, came home, packed my bags, and went back up.

Jane was constantly seeking the perfect relationship with her coaches. Her first two coaches disappointed her, left her feeling miserable and abandoned, and now she moved to a new environment, with hope to win the love and affection of her new coaches, the next targets of Jane's needs and desires. When the need for love and attention is not met by disappointing parental figures, sometimes those desires get projected onto a larger and more global stage (Australia will love her; the world will love her). There are also elements of this globalization in my story in chapter 10.

Jane's excitement about the training environment was short-lived, however, as she began to experience increasing pressure to perform to high expectations.

J: I stayed at the institute for quite a while, until realizing that it was not the perfect environment. I didn't know it at the time, but that's what they do. They are very cunning. The coaches are so nice. They allow for injuries, and you don't have to train if you're tired. "No problem. We'll cater for your needs." Yeah, that only lasts till they get you up there and then maybe for the first 2 weeks. Once you're permanent at the institute, they're nice, and then it starts to, you know, crunch down, but, um, I guess that's the coaches' way; you didn't really have a say.

Jane described feeling tricked by her coaches at the institute. She had hoped for perfection. Perhaps she had hoped that someone would care for her in the way she so desired. Someone might love her without her having to train excessively to win that love. Instead of feeling loved, however, Jane started to feel pressured, started to feel the demands to push herself.

I paused for a moment, wondering whether the coaches at the institute were responsible for putting pressure on Jane to overtrain, to "crunch down," or whether it was Jane putting the pressure on herself. Maybe her perceptions of the coaches' expectations were colored by her experiences with Matt and Jack, as well as her own internalized overtraining mentality. I asked Jane about the sources of these pressures. Her initial responses to my questions suggested that, not only did Jane tend to push herself, regardless of what others said or did, but also that the coaches maintained steady reinforcement for her overtraining behaviors.

J: I think I had a lot of that in me. When I was hurting, I did not talk about it, and I kept training through a lot, no matter what other people might say. But the coaches were putting more and more pressure on me as well. You know, if you did something wrong in training, they'd make you do more, or they didn't maybe realize you were tired, or sick, or injured. They didn't take that into consideration. They were like, "More, more, more is better than recovery." So we never got to recover properly. With injuries, we were actively getting treatment

> **" ...if you did something wrong in training, they'd make you do more, or they didn't maybe realize you were tired, or sick, or injured. They didn't take that into consideration. "**

and could still do something. We wouldn't rest; we would keep going. The injuries would never heal properly, and we would just have niggling injuries forever. Yeah, still a lot of pressure to perform.

With respect to her coaches' reinforcement for overtraining, Jane also pointed out that her institute coaches came from a different country, where cycling had a history of tough traditions that her coaches upheld and that athletes did not question. Those traditions from the coaches' homelands were similar to what Trisha Leahy talked about in chapter 6: Thrash and physically stress your athletes to the maximum, and those who are left standing are the ones you keep. Surrounded by these cultural traditions, Jane felt even more pressure to overtrain.

J: Okay, in their country, with these coaches, you didn't tell them what you can and can't do. They tell you what to do. It's the way they do things over there. They have a very militant style approach. They are very strong. They have always done things the hard way. If I gave them the information that my calf was not right, my coach, he would say, "Get out there and do it!" I ended up having knots in my calf that were going into spasm, but still out there, and still that's the way they do things in their country. To say, "No, I won't do that" is showing total disrespect; you just don't do it in their culture. Besides, I didn't have developed skills in saying, "I'm a bit too tired to do that," or, "I need a bit more recovery from that," or, "That is way too much for me to do." Coaches require you to do certain things. What happened overseas was one day we were about to go home, and I hadn't done a whole lot of training in that whole 4-week period because of the injury, but this one day we were at this track, and the coach wanted me to race against a couple of girls that this other coach had brought specially in. The risk of reinjury was huge. Huge! Two to three months after this was the trial for the Olympics, which I was hoping to compete in, and, in effect, I didn't. It was, literally, 12 months out from the Olympic Games, and, 12 months out from the Olympics, you don't want to flare up an injury that is going to put you off for at least another 3 months. So it's a really critical time thing, and it was ego; it was pride. It's the way they do things over there. In terms of the coach coming from that background, oh, how would you describe it for him if I did not race? Humiliating? I had no choice. I had to race.

Although Jane had strong tendencies to overtrain, even without a push from others, it seemed that she did not have much say in dealing effectively with her injuries. The way the coaches made Jane feel about injuries, and the way she felt pressured to train through them, was familiar territory for her. Jane had learned from a young age to keep silent about injuries, especially if she wanted to receive love and attention from her coaches.

While training at the institute, Jane never quite found what she was looking for in a coach. Even though she pushed herself hard, and tried to please her coaches, she never got close to them emotionally. Jane attributed her dissatisfaction with training at the institute to personality differences with her coaches.

J: A lot of it, I think, was my personality difference with the coaches. I didn't really get on that well with the head coach, and he didn't quite understand me as a person, and I'd always put a lot of pressure on myself, and then get angry and upset, and he'd kind of say, "Oh, too-hard basket!" [i.e., too hard to deal with] about me. We didn't click. With the training, I think I was just getting assembly-line kind of programs. It wasn't tailored for me. It was because at the institute, with so many people he had under him, and he had other big-name athletes, basically I was just getting their programs, and I'm not the same as them, so it didn't work. The cycle didn't work. I got pissed off, and we didn't get on.

It sounded like the coaches were not satisfying Jane's needs for them to take care of her. They were emotionally distant, yet demanding. Jane wanted the coaches to see her "as a person," but she felt that she was an object on an assembly line. She spoke of her training not being "tailored for me," but it seemed the lack of tailoring was also felt at the interpersonal level. She did not get along well with her coaches, and she was left confused about how to please them. At least with Matt, and especially with Jack, her mission had been clear: "If I train harder, push through injury, and don't complain about pain or fatigue, people will love me."

THE GOOD COACH CAN'T FIX HER

At age 15, Jane left the institute and gave up her scholarship because of a difficult struggle with overtraining syndrome, or what doctors had diagnosed as chronic fatigue. After almost a year away from cycling and any other type of training, Jane began cycling again in a new city with her new coach, Aaron. It was an opportunity for Jane to learn from her overtraining mistakes. Aaron promoted a balanced approach to training and recovery, encouraging Jane to work on her nutrition and her recovery and to take care of her injuries.

J: He knew that I had a really bad year, a really rough year, and we sat down and discussed things we could do differently, like going to a dietician, and having someone tell me about eating carbs. He told me that I would have to eat, sleep, and recover properly, and he would say that I should not be out there if I was injured. So, with Aaron, I went through the roof in terms of performance. I had big breakthroughs, really big breaks, and I just missed Sydney. Everything was going really well, and I was quite happy and settled.

From these first comments, it sounded like Jane might have found a situation that allowed her to change her overtraining habits. Aaron helped her to get to new levels of performance without overtraining her. It seemed almost too good to be true, and I was left wondering when Jane's internal slave driver would surface. In her next breath, Jane admitted that she did not always follow Aaron's advice. She acknowledged that she had internalized the overtraining mentality of her previous coaches, and she identified with that mentality, saying it was part of her. Jane admitted to doing more than Aaron would ask of her.

J: I am a bit of an obsessive kind of control person. I am pretty highly disciplined, pretty driven, and, when someone was pouring overtraining into me, like with Jack, I'd feed off it as well. But with Jack I had an excuse because I was quite naïve in terms of what was right, but now I am still like that; it's in me. Aaron could tell me not to do something, and

I'll still be driven to go out and do it, and ride extra kilometers at night, even if I was feeling tired. So, yeah, that is my nature for sure, a bit of an extreme person.

> **" He used to say, if your injury is distracting you and you could use it as an excuse for not doing well, then you shouldn't be out there. Now, unfortunately, the way I took that was to disassociate. Don't let it distract you, no matter what it is. "**

Overtraining had been so deeply ingrained in Jane's psyche as a way to win attention, fill her emptiness, and validate her worthiness that she kept resorting to those behaviors with Aaron, often against his suggestions to take more recovery time. It seemed odd that she would go against Aaron's advice, but in a way, it fit: "I have got this great coach who takes care of me. He is fantastic. Now if I can just do a bit more, and show him how good I am, then he will be really, really pleased." Though she knew, at some level, that what she was doing could be damaging, it did not matter. In the realms of desire and love, humans are often irrational beings and can simultaneously hold contradictory thoughts and emotions. Furthermore, Jane reframed Aaron's comments about injury in ways that kept her in overtraining mode.

J: My coach was very good about injuries. He would tell me to take a break, if I needed it. He used to say, if your injury is distracting you and you could use it as an excuse for not doing well, then you shouldn't be out there. Now, unfortunately, the way I took that was to disassociate. Don't let it distract you, no matter what it is. So, if I was hurting, I would be saying, "I'm fine," and would go and train when I'm dead tired, or I haven't eaten, or exhausted, or feeling sick. If I'd ever said how I was really feeling, he'd say, "Well, go home and rest." I'd say, "No, I don't want to go home; I want to ride. That is why I am here." So, I probably was a little dysfunctional in my way of dealing with that.

By this point in the interview, I could not help feeling frustrated that Jane had kept pushing herself under Aaron's leadership. I suppose that frustration came from my identification with Jane's strong internal drive to overtrain, coupled with envy that she had found a coach like Aaron, something I never did in my years as a competitive athlete. Why did Jane keep overdoing it? She was achieving success with Aaron. She did not need to push herself harder. Was the influence of her former coaches so strong that she could not change her maladaptive behaviors? I began to wonder if Jane's perceptions of how she might win love and approval from Aaron had something to do with her continued overtraining behaviors.

In the past, Jane had been accustomed to keeping silent about injuries and fatigue, and she had been rewarded for such silence. Perhaps she was looking to be rewarded with love and praise from Aaron in the only ways that she had been taught. Lessons learned early often become solidified and then operate by default. Even with a benevolent coach, who was trying to help her overcome her tendencies to thrash herself, Jane reverted to default mode. With abused children, even when they are placed in a caring and nurturing environment, we see a repetition of maladaptive behaviors that have their roots in early abuse experiences.

J: I'm really good at hiding how much I am struggling, so Aaron probably hadn't had the opportunity to say, "That's a bad example." They only see you as a good example because, when you come to training, you are *there*, mentally and physically. You know, if training is an hour and a half, you are there 20 minutes before you start. So, they use you as an example. That kind of pushed me even more because you want to be an example.

I sensed that wanting "to be an example" was equivalent to wanting *to be loved*, or respected and maybe revered, and wanting to be special. Jane believed that she had to keep up her image of being tough, if she was to win love and respect from Aaron. But wanting to be "an example" for other athletes suggests the globalization of needs that we spoke about earlier. It's not just for Aaron; it's for other athletes as well, and maybe she could even be an example for the world.

It sounded like Jane did not give Aaron much of a chance to understand what was going on with her body. She chose not to tell him when she was tired or hurting, which denied him the chance to suggest more adaptive responses to training and recovery. It seemed that Aaron was providing plenty of attention and caring for her as a person, but like trying to fill a void, it was not enough. The illogical reasoning might go something like this: "Aaron is wonderful, but I still feel empty.

Maybe I will be worthy and whole when the world loves me."

Despite having great success when she first started training with Aaron, Jane started to slip back into a state of fatigue. When I asked her about how her coach responded to this period of overtraining syndrome, she suggested that he could have been more available to support her.

SR: How was your coach around all these things? Did he really know what was going on?

J: Yeah, but he didn't have the time, really, to do anything about it. He was concerned and, obviously, he was telling me all the way along, "Jane, if you don't stop this, if you don't change this, and sleep, and eat properly, you're not going to perform!" I was kind of like, "Oh, whatever. I can still do it." Then, when I got overseas, I was just dead. I was just a corpse out there, and I thought, "God, it's true!" But a lot of me still had a lot of resentment in terms of, "You aren't there for me the way you used to be at the beginning. We don't talk, or have the time that we used to." I had all this pissed-off energy, and he'd say "Hi," and I'd say "Hi," and then just go and do my stuff, when inside I am thinking, "Actually, I am quite pissed off that I've come to this city because of you, and because of cycling, and I'm lonely."

Wow! All of a sudden, the anger was coming out toward Aaron. Jane was pissed off because he could not protect her from feeling lonely. Aaron was not fulfilling Jane's fantasized ideal of a coach, the ideal of a caring and loving parental figure, the one who *should* make everything better, and she was infuriated.

INTERNAL AND CULTURAL DRIVERS OF OVERTRAINING

After listening to Jane talk for over 90 minutes about how she had learned to overtrain to please her coaches and her mother and to emulate her father, and after hearing about her terrible injuries and battles with chronic fatigue, I was curious about how she might now regard her views on overtraining. When I asked Jane to reflect on her history of damaging behaviors, she admitted that she still struggled with her internal drive to do more: "I've got a lot of that in me now, where I'll train through stuff that I know most people would miss out on, even though I know better. So that's still in me."

Jane had overtrained and had admitted that it had been extremely damaging, physically and emotionally. Nonetheless, she still appeared to feel exalted in her tough image. It seemed that she still wanted to put herself above others for enduring more than they could. For example, she criticized fellow athletes for taking time off, complaining about being hurt, and not pushing through injuries.

J: I just see everyone else in my squad now just taking a day off because they've got a sore calf, and I am like, "Are you kidding? I'm competing with stress fractures!" I've known one guy, and he's really well-balanced as far as his body goes. I remember he had a fall, and I had to take him to hospital. It was embarrassing because he was squeaking so loudly, but he just says he has no problem about it. Now he's an incredibly selfish, self-centered, egocentric person. The universe revolves around him, and, I think, in a lot of ways, it works really well because he doesn't care who thinks he might be a wimp. If he's in pain, he's in pain. He lies down; he dies right there, no matter what other responsibilities he may have towards other people.

Jane equated taking time off for recovery, or complaining about injury, with being irresponsible and selfish. She distanced herself from an athlete who was more balanced in his approach to recovery, maybe because what he represented threatened her identity. But there was also a note of admiration for this athlete. It seemed that Jane wished she too could "not care about who thinks she's a wimp." Even though Jane criticized the athlete, I felt that she wanted to be as honest as he was. When he was in pain, others knew it. When Jane was in pain, she hid it. When the athlete was completely spent, he would "die" on the spot. When Jane felt exhausted, she just kept going. It sounded like Jane still had a distance to go in changing her thoughts about overtraining and freeing herself from her past. She had a model of an athlete in front of her that she thought "worked really well", but she could not shake her long-established patterns of behavior and emulate that model.

When I asked Jane about her perspectives on having a more balanced training program with Aaron, she admitted that she did not think she was training hard enough.

J: It seemed that the new training program probably wasn't hard enough. It was like I went from one extreme to another, and it was hard in terms of elite level. It was also quality technical advice that I'd never had before, but I'm so used to this kind of aerobic fitness and strong, general kind of training. I was used to doing more.

This comment illustrated the power of Jane's overtraining mentality. She had a new coach who had taken her to new heights of performance, yet she questioned the training program because she did not feel like she was working hard enough. Jane also talked about having a hard time taking a break after a recent injury, even though she had already gone through a crippling struggle with chronic fatigue, during which she had not been able to train at all.

J: Just recently, I got forced to have 4 weeks off when I came back from the Goodwill Games, and it was the longest 4 weeks. Instead of just embracing it and saying, "This is great. I need this. I haven't had this much time off in 3 years," I was fighting my coach, saying "Can't I just do something?" Now I'm back training, and I guess I shouldn't have whined. I find it really hard to sit back on a Sunday and go, "It feels good not to go to the track today." I find it really hard.

> **" ...there's an implication that, unless you're almost dead, you keep going. "**

Even after Jane's claims that she recognized her behaviors as maladaptive, she continued to overtrain. I asked her if there was an explanation for these contradictions between her thoughts and her behaviors. Jane struggled to find an adequate explanation. Looking back at her response to the onset of her chronic fatigue, Jane remarked that all she could think of was that she had not been fit enough, and probably had not been training hard enough.

J: I was used to being tired, but I wasn't going fast, so I mustn't have been training hard enough. So, I went out and did more training, and when I didn't complete a session, I'd come down in the afternoon, and try to complete a session and a half. It got to the point where, for a few weeks, I would go out and warm up for only a short time, and I would have to just stop, and turn around, and go home. Then I'd be so irritated with myself, and my performance, and that's when I had to make up for two sessions. It was just so ridiculous. I see it now, and I'm frustrated because I didn't recognize it then. I think I really felt like I should have been coping, and I was just angry about not coping as well as I wanted to, but I was very reluctant to take time off. I was used to completing all my training program all the time. Gosh, retrospectively how stupid was that? It was so silly because I wasn't being lazy, but, at the time, oh, I don't know, I just thought I was lazy. I thought it wasn't the program. It was my laziness.

> **" ...I just don't understand how these coaches get to where they are without a brain. Why aren't they there to help athletes? "**

The harsh internalized critic that Jane had developed from the influences of her mother, her coaches, and her father would not let her rest, even when she could see that it was "ridiculous" to keep going. That internal critic had kept telling Jane that, when she felt tired, it must be due to laziness.

If pressure from coaches, and from having her parents' attention and affection seem contingent on her sport success, were not enough to push Jane to constant overtraining, she also felt pressure from a tough sport culture. She said that athletes in her sport would be severely judged if they did not train through injury or keep silent about pains, niggles, and illnesses.

J: It's the culture of our sport as well. I think there are a number of things. It's very much, "Do or die. Don't be a wimp. You fall, you get back up." I didn't actively think, "I am not going to let it affect me." Instead, I was so in the culture that I just didn't let it into my head at all. That probably was quite a strong factor in why I didn't want to stop, because, for me, maybe I would have seen that as weak myself. You know, you break your arm; I've competed with a broken arm before. You suck it up. You keep going, and anybody that stops for a physical reason, there's an implication that, unless you're almost dead, you keep going. So, yeah, it's very much part of the culture. You'd better be dead before you don't go. It's a real tough sort of sport and to be sitting out at training was just not on. There was this mentality that you were weak, or you were a bludger [i.e., loafer or shirker].

Jane had heard the message that she had to keep training unless she was, figuratively, dead. I paused for a moment, thinking about this cultural imperative. If you are not "dead," then you have to keep training. The corollary to this idea was that if you were "dead," then you would not have to train, or at least you would not be expected to perform to expectations. Perhaps, then, if Jane *killed* herself with overtraining, she would have an excuse for not performing well. I started thinking that maybe part of Jane's overtraining had also been about self-handicapping. If she overtrained to the point of incapacitation, then she would have an excuse for not performing well. When athletes have overtrained, no one can accuse them of not working hard enough, and no one can judge them based on their performances because they were not at peak fitness. It seemed that Jane had been trapped, using overtraining both as a mechanism to win love and approval and as an excuse for failing to perform. She seemed to have a mass of irrational thoughts and contradictory motivations for training so hard. She knew, at some level, that she was probably training too hard, but still she continued. She trained to get respect and approval from others in her subculture, but she did so in a way that ensured she would not get the further admiration she desired because she had "killed" herself with her extra work. There also may have been a love–hate dynamic in regard to her mother. She would train hard to please her mother, and maybe to fix her mother's pain and disappointment, but then she would train extrahard to ensure she would fail, thus getting some sort of revenge (hurting Mum and reminding her of her own disappointments) for not loving her as she should have. As much as I resisted acknowledging it, I identified with Jane. Perhaps I, too, had used overtraining as an excuse for never making it to the Olympics in rowing. I also used sport to try to fix my dad's disappointments. Jane acknowledged that overtraining could be a trap, a vicious cycle, and I had never seen someone so trapped in such a self-defeating world of irrational thoughts, insatiable emotional needs, and damaging behaviors.

J: It was a whole cycle, and I was training so hard for the wrong reasons. It was also because I love it, and it's, essentially, why I'm here. I didn't want to feel that I had any reason not to train. So I fell into that trap, obviously, of just forgetting, or ignoring, that I was tired, or sick, or run down, and tried to forget about it, and just kept training.

As the interview came to an end, Jane left me with a final glimpse of her damaged emotional state. It echoed, for me, John's last words about wanting sport to "fix" him, because Jane also was desperate to be fixed. Unlike John, however, she did not look back at her career with regret. Rather, she looked back with anger that no one could fix her. Jane could not rely on others, so she relied on herself. Unfortunately, she had only one "cure" for her condition—to train more—and that cure for her malady was killing her.

J: There's certainly a lot of things that I have gained from my experiences, which isn't to say I'm not really very angry about all of it, still, and pissed off about the whole situation. I mean, I just don't understand how these coaches get to where they are without a brain. Why aren't they there to help athletes? It frustrates me a little bit that they see young athletes doing the wrong things, and they don't say, "You shouldn't keep going." But there's no benefit in me being wound up, or bitter, or twisted, or anything like that.

REFLECTIONS ON JANE

Listening to Jane talk about being "pissed off," I got the sense that she was angry at everything to do with sport, including the people in it. Like John, Jane was faced with the reality that nobody could "fix" her. Nobody could make her feel whole, or satisfy her desire to feel loved, or fill the gap that had not been adequately occupied even by her parents. It seemed to me that there *was* benefit for Jane in telling me that she was left "bitter, or twisted," from her experiences in competitive sport. Perhaps she was looking for validation of this maelstrom of negative emotions, which had been gathering in her over the years. In some ways, I wanted to give Jane this validation, to help her openly express her anger at competitive sport, but in other ways, I recoiled from the thought of criticizing competitive sport more than I already had. Jane also seemed to approach and then withdraw from her negative emotions, expressing them and then telling me that there was no benefit in allowing them to

continue within her. It is possible that Jane and I were colluding in protecting our now-tainted faith in the goodness of competitive sport. Steve and John felt regret about their experiences in sport; if they could just go back and do it again, then things would be better. Jane felt no longing to go back and change things; she was primarily angry. Jane started her athletic life off balance, with a heavy push from her mother, who was anxious to relive sport glory vicariously through her daughter. With her mother's push, and her father's modeling of overtraining, Jane did not have space to develop a balanced identity. She became a *perfect* candidate to develop overtraining, and most of her coaches seemed more than eager to support this development. When Jane finally did fall under the guidance of a balanced coach, she struggled against herself, finding it extremely difficult to change her pathogenic default patterns of behavior.

Jane illustrates in such a painful way how complex, contradictory, irrational, confusing, destructive, and imprisoning the world of an overtraining-prone athlete can be. That world is situated within a culture that encourages overtraining through the powerful emotion of shame. Jane fell into the trap of those external cultural imperatives. Her perfectionism—with its compensatory foundations in early experiences of being imperfect, seriously ill, and helpless—fueled her overtraining behaviors. Jane had features of obsessive-compulsive anxiety disorder, but it seems that her compulsions (training) served not so much to stave off anxiety as to fill that love-starved void. Maybe it was not the compulsive training that would keep anxiety at bay, but the results of that compulsion (love and respect from others) that counteracted the dread and anxiety she felt when she gazed into the essential emptiness of her being. Even more than John, Jane seemed like a "hungry ghost" (see the end of chapter 8). What she used to fill and nourish her emptiness was the same thing that caused her great pain and anguish.

After finishing the interview with Jane, I felt unsure whether she would ever change, and maybe I also wondered if I would ever change. I wondered if Jane would ever come to accept that she was okay without having to damage her body with overtraining. I left hoping that one day Jane would feel worthy of love, regardless of success in sport. I hoped that she would not have to use sport to distract herself from loneliness. I hoped that she might feel at peace with letting go of her sport when it came time to end her career. I hoped that one day the culture of competitive sport and the people in it, including coaches and parents, would cease to drive young athletes like Jane to overtrain. I hoped that both she and I could have our faith in the goodness of competitive sport restored. These hopes were big items for me and for Jane, but change has to start somewhere, and hope seems as good a place as any.

10

THE PERFECT BOY
The Author's Tale

In this final chapter of part III, Tony and Mark asked Sean to write some more about himself, to tell his story. We warned him that he might have to get relatively naked in front of the readers and reveal his frailties. Standing before an audience, warts and all, is a daunting task, but Sean has cast aside, for the most part, his trepidations, and he has added another layer to our exploration of the complexities surrounding overtraining (OT). Here is his tale told from his viewpoint. At the end of this chapter, Sean sums up and discusses all the stories in this part of the book.

SEAN'S TALE

Reflecting on Jane's story, it seems that her perfectionist behaviors and subsequent overtraining derived from experiences with an abusive coach and a demanding and difficult mother. Why, then, is this chapter about me titled "The Perfect Boy"? I am not like Jane. My experiences are totally different. It makes me cringe to be connected and compared with her through the word "perfect." I feel that my path, in sport and in life, has been in complete contrast to the path of someone like Jane; I do not deserve to be compared to her. Well, really, I feel that my coaches and my parents do not deserve to be compared to Jane's coaches or parents; in my mind, they were different from Jane's, better. They were, and are still, loving. They were supportive. They were the opposite of anything abusive. Yet here I am telling my story of overtraining and injury. Something must have

happened to thrust me into the same sorts of crises faced by Jane and the other athletes described in this book.

Unlike Jane, I felt I did not fit the overtraining identity all that well. I did not believe in the more-is-better philosophy; rather, I looked forward to rest days. I thought a lot about, and attempted to maximize, recovery. If I were injured or tired, I tried to take the best care of myself possible. Yet I still ended up with injuries that appeared to be linked to overtraining. My experiences led me to the key question of this book: What drives athletes to overtraining behaviors and outcomes in the first place? To answer this question, I thought I would ask people in elite sport for their perspectives and stories. For a long time, I did not identify my own story as associated with overtraining. I have, however, changed my views since beginning this project. I know now that, like so many athletes, I pushed too hard at the wrong times, blocked out niggles that turned into injuries, mismanaged my recovery, and was motivated to pursue unrealistic goals. So, in many ways, I do belong in this section of the book, and here is my story, one that is more about my unconscious needs to be loved than about overtraining, but it is the tale of what drove me to my limits in sport.

A Sport Fanatic From Early On

I love competitive sport. I always have, and I believe I always will. I started with soccer at age 6 and ski racing at 9. I dreamed of being a downhill champion, like one of the Crazy Canucks (Canadian

World Cup skiers of the 1980s). In high school, I began rowing, following in the footsteps of my father, a Canadian Olympic rower. At the same time, I took up competitive windsurfing, and my dreams shifted to being a top professional windsurfer. I wanted to be the best in the world, and I believed that I could be (I had my dad and his positive reinforcement to thank for my self-belief). In my mind, I constantly saw images of myself standing on top of a podium, bursting inside with elation and pride for being the best in the world. I was fanatical about windsurfing; I had plans to make it my life, and I trained hard to make that happen. At the same time, I continued in competitive rowing (hanging onto the possibility of rowing at the Olympic level if windsurfing did not work out). When I was 19, I had a shot at the professional windsurfing circuit with 3 months of competition in the Caribbean. I believed I would make a major, and immediate, impact on the pro tour, but despite encouraging results overall, I realized it would take much more time than I had imagined to crack into the top ten in the world. I still believed I could make it to the top in windsurfing, if I committed my life to it, but my short time on the tour convinced me that I did not want to do so. Touring week in and week out was not as appealing as I had imagined. Furthermore, a career as a professional windsurfer meant giving up any academic pursuits, which were important to me (and to my family). Rowing, on the other hand, was another sport in which I had the utmost confidence to succeed, and I believed it might fit better with what I wanted to do with my life, which included tertiary studies [university bachelor and higher degrees]. Perhaps it was also the most certain way for me to succeed in living up to my father's legacy as an Olympian.

Reflecting on my path and my desires in that stage of my life, I realized that the critical thing for me was to be the best in the world in whichever sport I chose. Where did that desire come from? It was a dream passed down to me from my father, a seemingly noble dream, but one that I now identify as tinged with sadness and disappointment as well—and not only my own sadness, but also the sadness, regret, and pain of my father, whose Olympic dreams were shattered by a sudden onset of pneumonia during his 1968 Olympic campaign in Mexico City. He has his own story, one which is intimately interwoven with my own. Rowing, for my dad, was a world that seemed to provide him with an identity, one that he cherished deep within his heart. He had grown up in a hardwork-ing family, without much time or opportunity to develop a social life, and rowing gave him a place to connect with others and feel good about himself. As with many sports, being on a competitive rowing crew can offer a refuge from the angst of human existence; it is an endeavor that seems to satisfy human needs for attachment, support, and camaraderie. The competitive environment and the possibility of success on the world stage also promise to provide the ego with fortifications to endure the inevitable attacks from many directions in life. I learned from an early age about the joy of sport, the social fun, the physical satisfaction, and many other salubrious effects. Nonetheless, I also developed a belief that I could fix my dad's pain through my achievements in sport. If I could win that gold medal or become number one in the world in my sport, and dedicate such an achievement to my dad and his support, love, and inspiration, then he would be fixed; I would have made up for his past pain. And I learned early on in sport that the way to get to the top is to work hard, to be mentally tough, to push through pain, and to persevere against the odds. Today, in my work as a psychologist, I see so many athletes trying to fix their parents and fix themselves through sport, and trying to do so by working harder and harder. I often recognize myself in the athletes I serve.

Soccer

I started playing soccer at 6 years old, and I had a coach who made it a lot of fun. My father came to games on weekends. I enjoyed winning. When we would win a game, there was a big celebration; I loved winning. I was in soccer for 7 years. Originally, I played on defense and did not appear to have a huge impact on the game. At around age 10, I moved into the position of goalie, following some training in taekwondo that made me more flexible and increased my quickness. For the next 2 years, I was the goalie. Our team won every single game, and I allowed only three goals. I was quite proud of my achievements, and I enjoyed winning the trophies and medals and being on the league championship team. I loved being part of a winning team. I don't remember my coach being a huge influence at the time; I just remember that he was great with the kids. He was very focused on skills, and he had a passion for soccer and loved to encourage us. He was a good coach. From a very young age, it was all about being involved in competitive sport, and Dad was always there supporting me.

Developing an Internal Critic

My supervisor (Mark Andersen) once asked me if I could recall when my harsh internal critic started to appear. The first thing that came to mind was ski racing. With soccer, I had a really good experience. When I became goalie, I felt like I was one of the most important people on the team, and it was a very successful team. It was all so positive. When I made the transition to ski racing, I was still playing soccer. At age 10 or 11, I remember watching Canadian Steve Podborski win the overall World Cup downhill title with a victory at Whistler in my home province of British Columbia. It was a time when the Crazy Canucks were considered the most daring downhill racers in the world. They were some of the first non-Europeans to break open the World Cup downhill scene. I started to develop dreams of being a downhill ski racer like the Crazy Canucks. If I could do that, then the world would know I was okay. I think the development of huge aspirations was accompanied by the growth of an internal critic that kept reminding me that I was not reaching those lofty goals.

I will digress here for a moment. Going back to when I was about 5 or 6 years old, my dad was still involved in rowing, competing on the Canadian national team. I used to go to the rowing practices and sit in the coach boat. I was around a father, from a young age, who was still a national-level athlete. My dad had always talked about rowing and his own failed Olympic dreams. We would watch sports together and would get excited about seeing things like downhill skiing. I recognize now that my identification with these great ski racers evolved from this idea of wanting to follow in my dad's footsteps and become a World or Olympic champion. My dreams came from my father. It also seemed, at the time, like a realistic dream, given that I grew up in this environment of competitive sport, with a father who was still competitive at a high level.

With ski racing, one generally trains both days on the weekend if one can, but at first I was still playing soccer on Saturdays. To start with, skiing was not the greatest experience. I was not a natural for the more technical events like slalom or giant slalom. I believed, however, that I would be the best at the downhill discipline because it was all about conquering fear, which I thought I could do, though kids do not compete in that event at a young age. One thing that sticks in my mind is a group of really good kids on the ski team, including my yet-to-be best friend. They used to taunt me for not being totally committed to ski racing.

I would be going through the racing course, and they would yell at me from the chair lift, "Go play soccer, Richardson." Thus I probably felt a little bit on the outside with the ski team. Perhaps skiing was the first experience of feeling not quite good enough. Nonetheless, in all my sports I have always had the belief that if I'm really committed to it and decide that this sport is the one that I want to do, I will succeed. I probably attribute that to the positive reinforcement from my father. He always reminded me that whatever I wanted to do, I could accomplish. He believed in me 100%, and he let me know that. He made me feel that I could accomplish whatever I put my mind to. I always believed him, and I still do.

Reflecting on it now as a psychologist, I recall that I've always felt the confidence to succeed in any endeavor, but motivation comes into play, and motivation and confidence are not the same thing. If I felt motivated enough to do something, then I would go after it and I would succeed. Yet there were times when I thought, "I know I can do that, but I'm not really that motivated." After I stopped playing soccer, I continued to ski-race for a few years to see if I could succeed at it. I still had the image in my mind of the Crazy Canucks, the ones who got the glory. They were the best in the world, and downhill was their discipline. I believed that I just had to get into the downhill, but teenagers usually do not compete in their first downhills until age 15. I was probably working toward the downhill, but at the same time I started windsurfing competitively, and I started rowing. I got into rowing because of my father, though he never pressured me to do it. I just thought, "I am going to do it." I still loved skiing because it was a fun sport.

Eventually I gave ski racing away, partly because I was growing quickly and had knee problems (Osgood-Schlatter disease) and partly because I did not want to do it anymore. At the beginning of my last ski racing season, my knees were very painful, so I ended up quitting the sport. Reflecting on that time, I see that, regardless of my knee problems, I was not totally committed to ski racing as a career. I loved skiing, but perhaps I did not love ski racing enough. Perhaps I was not getting good enough, quickly enough. I do remember thinking that the injury was a nice excuse for my lack of successful performance. I was not unhappy to quit ski racing, and it was nice to have something to point to, like a growth-related injury.

My internal critic is probably connected to, and developed along with, my idealized image of myself (world number one) and my idolization of sport stars like the Crazy Canucks. In thinking now about windsurfing, which was becoming my predominant sport, I completely idolized Robby Naish, the top windsurfer in the world. I began to realize it was possible to compete in windsurfing when I visited Hood River, Oregon, to train and to watch professional international racing events at which most of my major idols competed. I began to develop images of becoming number one in the world in windsurfing, and I started competing at the junior level. I loved windsurfing more than any other sport, and I was excited at the chance to combine my passion for the sport with the possibility of competing at a high level. My drive to succeed in windsurfing also supplanted my fading dreams of ski racing. Rowing was still in the background, but I did not have the same passion for it as for windsurfing. In the back of my mind, I always thought I could still get into rowing at a later age because Dad was an Olympic rower, and he let me know that I could do it whenever I wanted. Rowing was in my genes.

Windsurfing took over my thoughts and my life from age 15 to age 20. I was a fanatic about it, and I had a couple of friends who were the same; all we talked about was windsurfing. One of my best friends and I planned to become professional windsurfers. I was also dedicated to training as much as I could. Looking back, I can see that I was disciplined, hardworking, and mentally strong in terms of my training and practice. I never experienced any sort of overtraining in windsurfing, but I was ready to work harder than others to get where I wanted to go. Perhaps the one thing I prided myself most on was my mental discipline—I practiced my imagery for hours, seeing myself succeed at the highest level, and I felt confident that I would get there. I think I was starting to develop the mechanisms of mental toughness that led me blindly to wreck myself later in life when rowing. I recall a funny moment when I was 16. I gave my girlfriend a picture of me windsurfing and wrote on the back: "Remember this face. One day I will be number one in the world!" It sounds arrogant now when I look at it, but I was a teenage male, and arrogance and cockiness come with the territory. I was so focused then on becoming number one in the world, and I had a lot of encouragement for that ambition. I had sponsors, and one of them said, "You have got what it takes. You can make it in this sport." So, when I finished high school, my idea was to have a crack at the pro circuit.

Where did that massive desire to be, in the true sense of the word, unique, one of a kind, number one in the world, come from? Dad. Without a doubt. I have written about him earlier in this book, so perhaps my answer is affected by what I have already thought about, but the desire feels very much connected to growing up with Dad's legacy of failed Olympic dreams. He talked a lot about how he so wanted to win a gold medal, and he was in a crew who could have succeeded, could have been number one in the world, except that he succumbed to pneumonia and almost died in Mexico.

I think Dad, however, is only part of the story. Thinking over what I wrote earlier about not quite making it in ski racing, I was feeling a bit on the outside. It is almost a sense of fulfilling my destiny in terms of *how I feel about myself*. My dad has told me my whole life that I can succeed at whatever I want to do, but maybe it is about my *believing in it myself*. Did I need to be number one to answer the part of me that questioned my dad's faith in my capability? Perhaps I needed to prove my dad right so that I could finally believe what he said to me. Looking back, I recall being aware that I was not the best in a number of sports that we played at school, such as basketball, rugby, or football. Nevertheless, in comparing myself with others, I would protect my ego by saying to myself, "He may be the best at basketball, but I am a better windsurfer." Then when it came to windsurfing, I would say, "He may be the best windsurfer, but I am better at rowing." And I could further ensure protection of my ego if I trained and pushed myself harder than others.

Maybe I will be okay with myself if I can prove to myself that I am number one. It is difficult to reflect on much of my history because I have changed how I think. What comes to mind is a feeling of being a kid and wanting to be noticed. When I was very young, I was quite shy. I was a jumble of contrasts because I was always enthusiastic and noisy within my own family, animated and excitable. Yet, in the social environment, I could be quite reserved and inhibited.

I was never the most popular kid in school, but at the same time I was not unpopular. I was in the middle. I think I managed my relationships with other kids well. In high school, I was happy because I had the identity of being a windsurfer. High school is tough on most kids, and I had developed an identity to protect me and to hold on to. I was not in the popular group, but I did have my own sense of self. I guess I was not distinguished from the crowd in high school. Socially, I was not unique, but I carved an identity as being unique, a windsurfer, and I was going to be number one in the world. I always thought that when I returned for my high school reunion, people (especially the girls) would say, "Wow, look at Sean. Look at what he has done with his life." I guess the fantasy about being world number one was the pathway to being loved, to being attractive and accepted. It's interesting that my dad's positive reinforcement, always saying, "You can do it," contrasted with my social reinforcement in the high school environment, where I was unconvinced that he was right. Underneath it all, I had the sense of, "I have got this in me; I know I can do it. I just have to show it to everyone else." My Dad sees me for who I am, but no one else does. But then, because no one else does, maybe I have to prove it to them, and prove it to myself. Connecting these reflections to my overtraining behavior in rowing, I imagine that I had found a sport where I might prove my dad right, and put aside my own doubts and feelings of being left out, and all I had to do was train harder than everyone else.

When I was 15 or 16 and really getting into windsurfing, my friends and I also formed a rowing crew composed of guys at our year level, with my dad as the coach. In that group, we were not at the top of the social ladder; we were in the middle. We were not the football players. We were all just good guys, and we ended up winning the equivalent of the State Championships. Nonetheless, I maintained my perspective on rowing as a means to getting fitter for windsurfing. In the back of my mind, I still viewed rowing as something I could do later in my life, as my father had done. For now, windsurfing was my focus. When I finished high school, I started to look toward competing on the professional windsurfing circuit. When I finished high school at 17 or 18 years old, I was aiming to win the amateur men's division of the Pro-Am International Slalom windsurfing competition in Oregon. I ended up second in my final year as an amateur because I got nervous and crashed while racing. I was the fastest but ended up second. I was still happy with it, and sponsors encouraged me. I went to university the next year and joined the rowing team. At 18 years old, I was the smallest on the crew, weighing 185 pounds at 6 feet 3 inches, whereas most of the crew members weighed more than 200 pounds and stood 6 feet 4 (or 5) inches. It was almost as if I got into rowing just to see what was happening in the sport, without really being passionate about it. I also saw it as training to get stronger for windsurfing. I planned to get a wildcard entry to the pro windsurfing circuit the next

summer, and rowing was a chance to get strong and fit over the winter while developing in a sport that I might pick up later. I ended up being the fastest rower in the heavyweight men's varsity program. I won every intracrew seat race, no matter which partner I had. One of my crew went to the Olympics the following year as a spare for the gold-medal-winning Canadian Olympic Eight. Reflecting on that year, I wondered, "Why did I not get recruited if I was so fast?" Perhaps my coach did not bother because he knew I was so focused on windsurfing. But if I was that fast, and I was so young, why were they not paying attention to me? This was perhaps one of my first experiences of regret or a sense of missed opportunity. I didn't see it like that at the time, or even after the Canadians won the Olympic gold. Upon later reflection, however, I did.

I received the wildcard entry to the pro windsurfing circuit, and I ended up racing in five pro events. In my mind I was going to burst onto the scene, and, in some ways, I did well to start with, finishing 34th overall among the top 80 guys in the world. Actually, after four rounds of racing I was in 28th place, but I got nervous, crashed, and slid back to 34th. I was, however, encouraged. It sounded like a glamorous lifestyle: traveling, staying in hotels, and windsurfing. I returned to the Pro-Am in Oregon in July and placed 22nd overall; nonetheless, I still got nervous and fell and did not perform to my potential. I was a bit overwhelmed, but I realized I had done well in my first pro competitions. I knew that if I put my mind to it, given my speed and how well I was doing in my first couple of events, I could get to the top of the world in slalom racing after a few years of competition. I could work my way up to the top 5 in the world if I really wanted to.

But I started not to want to. I came back from the windsurfing and thought, "To succeed here, you have to move to Hawaii, and although that sounds great, you have to give up education." Basically, you become a professional windsurfing bum, and you travel every couple of weeks around the world. I gave up on the sport, and on my youthful ambitions to be a pro windsurfer. It was a lot different from ski racing, because I actually got to race on the pro circuit and had won some prize money, but it was not what I thought it would be. In ski racing, I never got to see what it was like. With windsurfing, I did. I wondered, however, if there was an internal switch that would make sure I didn't get where I wanted to go.

Perhaps fear of failure was one of my concerns, but I don't think I was afraid of not making it in windsurfing. I can see the connection to my experiences in skiing, theoretically, that deep down I might have been scared, but it doesn't feel like that. I'm thinking that maybe it was more about what I was chasing, being unique, being number one in the world. In experiencing the pro circuit, I realized it was not going to provide what I was looking for. I was still not able to fill the gap, a gap that I did not quite understand back then. That gap, which I would describe now as the emptiness human beings feel when they realize they are alone in this world, I think comes from unfulfilled desire to be loved unconditionally. My parents would say they loved me unconditionally, but I, like most others I have counseled, probably doubted the "unconditionality" of that love at some level. Somewhere in my development, probably as a young child, I had experiences which I interpreted (rightly or wrongly) to mean that I was not good enough and not worthy of unconditional love. I must also have created a story in which being number one in the world would be the way to resolve this worry.

I think my dad was a big influence in terms of the mental side of performance, and I always felt mentally strong in my sports. I was especially strong in rowing, where my father had modeled mentally tough behavior, demonstrating and sharing how to cope with the pain and maintain the effort required in the sport. It was easy for me to be strong in rowing because I had learned at a young age to master the pain and push my body hard. Rowing was about being mentally tough, and my dad had taught me how to do that.

I returned to competitive rowing during my master's degree. I probably found my real passion for rowing when I switched from crew boats to rowing in the single. In crew boats, one depends upon other people, whereas in the single it is all about individual effort (and I suppose it was only in a single that I could take all the credit for success and prove myself truly worthy of unconditional love and admiration). Without expecting it, I immediately experienced success. I ended up as national champion in the university varsity single and was invited to the national team camp a year and a half before the Olympics. In my mind, there was no doubt I would make the Olympic team. By January 1999, I had been training seriously for 2 years and was ready to compete against the best in the country. I placed well at an unofficial national time trial, not too far behind Canada's former world champion single sculler, Derek Porter, an absolute god in the sport. I believed I could beat him later that year. I had moved from just having idols (e.g., the Crazy Canucks) to having idols that I wanted to knock off.

Reflections on My Story

My desire to be special and unique intersected with the world of rowing. Part of my story of going for Olympic gold is related in this book's introduction. My needs and the challenge of rowing for Canada at the Olympics combined in ways that led me to damage myself (like John) in pursuit of a place on the 2000 and 2004 teams. My 2004 pursuit was the most intriguing in that I was armed to the teeth with knowledge about overtraining, but I succumbed to the pressures to train while in pain, to try to "prove" myself to teammates and coaches. I saw what I was doing; I knew that I would damage myself, yet I continued. My knowledge of overtraining was not helpful. As one of my supervisors says, "All the knowledge in the world can't compete with the human heart." My heart's desire was to be an Olympic champion, to be unique, to be special, to finally fill a void. I have realized that the void can't be filled, and that perhaps part of my journey in training as a psychologist who works with performers is to use my experiences in the service of others.

When working with athletes, I question where I am in the process. What do I get out of the interaction with athletes? How does helping them affect me? I receive feedback that what I do with athletes is not the same as many other sport psychologists. I guess what I sell to athletes is that if you're okay with yourself, you will be all right, and ultimately you will perform. And maybe I am trying to sell that message to myself as well. I am aware that the messages I send to other people are about balance and being okay with yourself, and taking the approach that if you allow yourself to be you, your performance will follow. When I sit down with an athlete, I do not say, "I am going to give you something to improve your performance," like inserting a chip into a computer and all of a sudden great performances come out. Rather, I try to help athletes remove the barriers to being themselves. I think there is definitely something in that for me. I reflect on it, and I realize that a lot of what I say to others has a lot to do with how I feel about myself. Maybe in treating others, I am treating myself. By convincing others to be themselves, I am trying to convince myself that it is okay to be Sean, because in some ways it is still really quite okay to be just Sean. Even as I write that it is okay to be Sean, however, I can sense my hesitancy in accepting

Some sports, like rowing, are about being mentally tough, and they're about mastering the pain and pushing your body hard. That kind of sport culture encourages overtraining.

that statement. Only recently have I reconnected with some of my long-buried feelings and realized that I have been harboring an unconscious fear of not being accepted by the people I care about most. In some ways, it sounds ludicrous to me even to question what my family or partner might feel for me, but the human psyche seems to create lunacy effortlessly. Only a month ago, I uncovered a seemingly harmless childhood memory of being lost in a shopping mall, and another of getting angry at my mother for accidentally spilling hot water on my arm at age 5. In both of those memories I recall being shocked, angry, and traumatized all at once—and I probably questioned my Mom's love, despite these events being simple accidents. Questioning love over a simple accident, however, is the emotional logic of a child. The unexpected thing about my fears is that I came upon these childhood memories while reflecting on the nervousness I used to experience while sitting at the starting line of a rowing race. Funny how our minds and hearts work that way.

I see this fear of not being accepted in my relationships with other people, especially my close relationships. Probably the most salient characteristic of my past relationships is that I did not allow myself to be me. I felt stifled. I tried to be there too much for other people. In trying to take care of others, like trying to be there for my mom, I forgot about myself, then got frustrated, angry,

and resentful. What is odd is that my mom was not particularly supportive of my sporting ambitions. She really did not like it because she saw me in pain in my sports, and she had seen my father suffer in his sports, both mentally and physically. She did not want to see me go through the same trauma. She saw my dad thrashed and seriously ill, and it scared her. She still offered me her love and support throughout my sport career, but she stood apart. Mom would probably have preferred to have a son who was not like her husband. So in my relationship with my mom, and in significant relationships with others, I have stifled myself to protect and help them but ended up frustrated and resentful at being unable to be myself. The paradox here is that I have been resentful toward those I love, but the actual cause of the problem has been my own behavior. By stifling myself, by hiding parts of me, I have denied those I love a chance to accept me for who I am. Humans are funny creatures, indeed.

I think sport became the arena in which I tried to express myself, to say to others, to my parents, and to the world, "This is who I am. This is me, and I am good. I am okay. You can accept me for who I am. I cannot be stifled anymore if I am number one in the world, and I am unique." The unfortunate thing for me is that in my quest to be me, to be recognized as unique, to be heard, to be noticed, to be accepted for who I am, I tried too hard. I became too *mentally tough* for my own good and found myself on the path to overtraining. I am now preaching balance to athletes so that they can feel good about being themselves in a world that is always asking more of them. I think that practicing balance is the healthiest way to be successful in sport (although it certainly has not been the only way, as many athletes continue to push themselves through injury, fatigue, and illness in pursuit of their goals). Perhaps this book, and the stories it contains (including my own), will help others follow healthier and more genuine paths to sport performance, paths that foster self-acceptance and excellence at the same time.

ALL OUR STORIES

Although I have often commented that I did not initially choose this research topic (I was asked to study overtraining by a university professor), so it is not really about me, I have come to acknowledge that I cannot separate it from my own experiences. Hence, my quest for self-understanding and self-acceptance has been a large part of the project's evolution, and it is intimately connected with the stories of the athletes in this book. Thus, I would like to reflect, sometimes in a personal manner, on all the stories told in this part of the book, and on how they relate to the literature we have covered and what the experts have had to say.

ATHLETES' PERSPECTIVES AND RISK FACTORS FOR OVERTRAINING

Reflecting on the aggregate stories of Steve, John, and Jane, and the 13 athletes on which these three stories are based, I have been fascinated, surprised, enlightened, saddened, and angered. Sometimes, the tales of these athletes flowed from my fingertips. At other times, I struggled, word by word, to present their stories and my interpretations of them. I have realized that the most difficult parts were those that keyed into my own pathology and resistances related to overtraining. Parts of their stories I reluctantly identified with, but I wanted to block those bits out, often engaging in defensive suppression, much like the athletes who were telling me their stories. At times, I got lost in the interviews, forgetting that the goal was to draw out risk factors for overtraining. At these moments, I might have feared that I was not asking the right

questions, those that would prompt athletes to tell me about OT risk factors. I realized, however, that following the athletes in the interviews was more important and would reveal more than pressing forward with my agenda about risk factors. Regardless, after constructing the three tales, I saw that more was exposed about risk factors, and about pathogenic behaviors by athletes, than I had anticipated. Supporting the experts' perspectives on OT discussed in chapter 4, the people I interviewed displayed personal characteristics of OT-susceptible athletes. They also experienced forces (e.g., sport cultural ones) that motivated them to train harder or push through injury, and they went through times where situational factors increased their stress loads and negatively affected their capabilities for adequate recovery. The athletes' stories went further, however, than the expert perspectives, revealing some of the less-talked-about issues surrounding overtraining—in particular, psychodynamic and other familial influences on athlete behaviors. The next two sections present what I feel are the most significant themes to emerge from the athlete interviews: the influences of coaches and parents, and the issue of injury within the context of overtraining. I then present a comparison of athlete perspectives on, and experiences with, overtraining from the interviews and from materials presented in chapters 2 and 3.

Coaches and Parents

The most salient OT risk factors emerging from the athletes' stories were the influences of coaches and parents. In examining both personal and situational risk factors for OT, it seemed that coaches' and parents' behaviors could be represented as both external and internal drivers. As external drivers, coaches and parents overtly pushed athletes to overtrain, telling them that the only way to succeed was to train harder. As internal drivers, representations (internal objects) of coaches and parents often became entrenched in the athletes' psyches, and in some cases entwined with their identities. Through patterns of reinforcement, and through contingency-based expressions of love and approval, coaches and parents provided the impetus for athletes to internalize overtraining approaches to sport. The athletes' stories suggested that they often started to develop overtraining behaviors at a young age, when they relied heavily on their coaches and parents for guidance and support. These early experiences seemed to be crucial in forming the athletes' responses to overtraining demands later in their careers. The athletes represented in Jane's tale illustrated the potential for these internalized experiences to drive athletes to overtrain even under the leadership of balanced coaches. With coaches who do not take a balanced approach, the risk for overtraining is always present. Even when athletes get older and might feel more comfortable speaking up for their needs, they remain susceptible to being pushed to overtraining by their coaches, because most sports are hierarchically structured, with the power allotted primarily to the coach. The influence of a coach operating in such a hierarchical structure was illustrated in the experiences of athletes who made up Steve's tale of professional football, and I imagine it might be replicated in many other professional and amateur sports. The athletes who became John illustrated the roles that coaches play in colluding with athletes in their pursuit of Olympic glory, in their desire to be worshiped as heroes, and in their belief that overtraining is the way to get to the top.

This emphasis on coaches' and parents' responsibilities for creating, reinforcing, and maintaining overtraining behaviors is critical to me. I can think of several examples of athletes I have known who have been blamed for their injuries and other overtraining outcomes. Coaches or teammates have said that the injured athletes brought misfortune upon themselves because they were perfectionists or because they did not take responsibility for their own bodies. In a way, it sounded like they were blaming the victims, and perhaps influencing the athletes to stay trapped in their cycles of misfortune. Human behavior does not occur in a vacuum, however, and past experiences tied to current realities and demands are significant contributors to the creation and maintenance of personal behavior patterns, dispositions, and traits. I sensed that these overtraining behaviors I witnessed in fellow athletes were largely products of their experiences in sport and of their experiences in their dynamic interactions within their families and coaches.

The overtraining behaviors described in the tales of Steve, John, and Jane (as well as my own) illustrate how overtraining habits are initially adopted to please coaches and parents. Child athletes most likely could not design complex training programs or understand the requirements for maximizing training and recovery in competitive sport, so they turn to their coaches and parents for guidance. If those coaches and parents promote overtraining behaviors, make expressions of love and approval contingent upon the young athletes' hard training efforts and sport successes, and do not teach the athletes to take care of their bodies, then young athletes seem to be left with no alternative but to adopt the more-is-better approach.

Injury in Overtraining

As mentioned at the outset of this book, I thought originally that I would be looking at overtraining processes and only one outcome—overtraining syndrome—when exploring OT risk factors. In the interviews, however, I found that one of the most common outcomes for athletes engaging in overtraining behaviors was injury. Furthermore, the athletes' experiences of injury, along with maladaptive responses to it, often served as the source of future overtraining behaviors. For example, many athletes who were used for all three stories experienced significant injuries, about which they felt guilty, as a result of which they were left feeling behind, and to which they reacted with increased training efforts, premature returns to training and competition, or further denials of the consequences of new injuries—all leading to worsening of the injuries. Thus the injuries and the overtraining responses to them created vicious cycles from which the athletes had difficulty escaping.

The issue of injury in the context of overtraining is not a novel concept; researchers (e.g., Kibler & Chandler, 1998) have talked about injury as a significant outcome of overtraining. Nonetheless, it seems that much of the OT research (e.g., Hooper, Mackinnon, Howard, Gordon, & Bachmann, 1995; Morgan, O'Connor, Sparling, & Pate, 1987; Rowbottom, Keast, Garciawebb, & Morton, 1997; Uusitalo, Huttunen, Hanin, Uusitalo, & Rusko, 1998) has been concentrated on one outcome—overtraining syndrome—and not on the process of overtraining, which can lead to various adverse outcomes, including injury, illness, OT syndrome, and numerous other physiological and psychological disturbances. My sense is that when talking about overtraining and doing research on OT syndrome, it might be important to specify that OT syndrome is only one possible outcome. The narrowness of focus and the general lack of acknowledgement of injury as a significant outcome of overtraining processes, or of stress–recovery imbalances, adds to the risk of misattributing injury to bad luck or to acute uncontrollable factors, when the causes might be rooted in controllable, chronic behaviors. The inclusion of injury in discussions of overtraining also has relevance to sports where overtraining syndrome is unlikely, but where, nonetheless, overtraining processes can occur. For example, in sports that emphasize technical training over fitness training, such as diving, athletes might not stress their bodies aerobically to the point where they experience the fatigue and exhaustion symptoms of OT syndrome, but they may overstress a particular muscle or joint from repetitive technique work and inadequate recovery of that muscle or joint. Furthermore, in all sports, there is potential for an overtraining response to an existing injury (i.e., trying to return to training too soon, thus increasing risk for reinjury).

Before leaving this section on injury, I think it is important to mention that the experience of illness in the context of overtraining shares similarities with that of injury, and it is also a significant outcome of overtraining processes. Although illness was not highlighted as much as injury within the athletes' stories told here, many of the 13 persons interviewed talked about struggling with viruses, colds, and other debilitating illnesses during times of peak training. In support of this link between overtraining and illness, Mackinnon (1998) pointed out that frequent illnesses are considered common outcomes or symptoms of overtraining, with similarities appearing between symptoms of overtraining syndrome and of infectious illness. Steinacker and Lehmann (2002) also suggested that high training loads are associated with increased risk of infections, and Nieman (1998) noted that endurance athletes are at higher risk for upper respiratory tract infections during periods of heavy training. Similar to the niggles experienced during the early onset of injury, minor colds and infections might be indicators of stress–recovery imbalance, which, if left unaddressed, could lead to more serious overtraining outcomes. Athletes may try to train through illnesses instead of taking the recovery necessary to get better, or try to return to training too early after having had time off due to illness.

ATHLETES' STORIES COMPARED WITH ATHLETES' EXPERIENCES FROM THE LITERATURE

To get a sense of how the athletes' stories in this part of the book reflect other athlete experiences presented in the literature, I have identified parallels between quotes from the athletes I interviewed and those quoted by a number of researchers in the field on several key issues:

- the roles of coaches,
- injury issues in overtraining,
- sport culture, and
- personal factors.

Coaches

In the literature, several authors (e.g., Kellmann, 2002; Krane, Greenleaf, & Snow, 1997; Wrisberg & Johnson, 2002) presented quotes from athletes illustrating the strong influence that coaches have in driving OT behaviors. In the examples presented in chapters 2 and 3, and in the three aggregate athlete tales in this part of the book, it is evident that coaches pushed athletes to overtrain through excessively demanding, often abusive, training practices; through patterns of reinforcement; and through sometimes subtle attitudinal influences. In particular, the 13 athletes represented in the stories of Steve, John, and Jane recounted experiences similar to ones presented by Kellmann (2002) and Krane et al. (1997), in which coaches pushed athletes too hard and the athletes followed the coaches' demands despite feeling that they were being overtrained.

The story that Krane et al. (1997) presented of a young gymnast driven by abusive coaches echoed the experiences of the melded athletes in Jane's story, especially those experiences with the coach we called Jack. These athletes appear to have been aware that their coaches were pushing excessively, but they continued to obey their coaches, believing it was their methods that brought them success. The athletes appear to have been so attached to any possibility of receiving love and approval from their coaches that they would do anything to keep the coaches happy. Similar to the contradictory emotions directed toward an abusive coach illustrated in Jane's story—in which the athlete shifted back and forth between love and hate—the gymnast in Krane et al. said that her coach was "excruciating" yet "wonderful" for pushing her to train with pain and to endure physical punishment for making mistakes. In my construction of Jane and in the story from Krane et al., the athletes appeared to have been trapped in dynamics with their coaches that replicated coercive, conditional dynamics that they transferred from their relationships with their parents.

With respect to coaches' abusive behaviors, there were also parallels between experiences detailed in Steve's and John's stories and experiences of athletes quoted in Wrisberg and Johnson (2002).

In particular, Wrisberg and Johnson presented quotes from an athlete who talked about feeling humiliated, denigrated, and verbally abused by her coaches. Athletes incorporated into both Steve's story and John's talked about the humiliation they suffered when coaches used them as illustrations for teammates of *bad* behavior or called them derogatory names. In all of these cases, the coaches' abusive behaviors prompted athletes to turn to overtraining as a coping mechanism. It seems that the role of the coach in influencing overtraining behaviors and associated adverse outcomes cannot be overemphasized. Furthermore, looking at the experiences highlighted in Jane's story, where athletes internalized their coaches' maladaptive attitudes and behaviors, it seems that one should take into consideration all of an athlete's previous experiences with coaches when assessing risk for overtraining.

Injury

Another significant theme to emerge from both the literature review and the athletes' stories in this book is the threat that injury mismanagement poses to athletes' well-being in the long term. Coaches' attitudes and behaviors appear to be the most influential factors affecting poor injury management. Both Wrisberg and Johnson (2002) and Krane et al. (1997) quoted athletes who had been pushed by their coaches to train and compete even though they were seriously injured. In the example from Krane et al., the young gymnast talked about training and competing with multiple serious injuries. In the experiences outlined in the stories of Steve, John, and Jane, athletes described being pressured by coaches to train or compete with stress fractures, pulled muscles, and broken bones. Coaches used abusive and sometimes coercive tactics to influence athletes to push through injuries, sometimes getting angry at athletes for being injured. At other times, they acted childishly to make athletes feel guilty for being injured. In all cases, coaches seemed to promote the delusion that injuries would go away if athletes blocked them out.

Sport Culture

Athletes in this book and in the literature talked about pressures they experienced from the subcultures of, and social expectations surrounding, their sport. Gould, Jackson, and Finch et al. (1993) provided an example of a figure skater driven to think obsessively about her weight by the constant emphasis placed on appearance in her sport. This obsession with weight, and the corresponding overtraining behaviors aimed at losing weight, were reflected in Jane's story of struggles with disordered eating in an attempt to meet the demands of her coaches and the perceived aesthetic demands of the sport. Steve's story also included an example of struggling to maintain an image imposed by the sport's subculture (represented in this constructed story as the image of the tough footballer).

In terms of living up to expectations of the general public or of people in sport institutions, Gould et al. (1993) and Wrisberg and Johnson (2002) provided examples where athletes expressed feelings of shame, of being condemned, or of being stressed-out for not living up to the standards set by those around them, including people from athletic organizations. Several of the athletes interviewed in this book talked about feeling shamed or humiliated for not living up to cultural imperatives. Athletes in Jane's tale talked about keeping silent about injury and maintaining facades of health because of the pressures they felt from people at elite sport institutes. In Steve's story, the pro footballer talked about the heavy burden of expectations from the coach, the team, the media, and the public. One of the Olympians in John's story felt "crucified" in front of the world for his failure to make his country proud. In all of these cases, the perceived pressure of expectations from others added to the stress loads that caused these athletes to need extra recovery. For some of them, the added pressures also led to pushing excessively in training in order to compensate for their perceived shortcomings.

This issue of cultural imperatives to push bodies to the point of severe damage and incapacitation has been largely left out of discussions on OT, but the problem has been examined in sport sociology research linking dominant codes of masculinity to sport injury risk (Young & White, 2000). In reviewing the literature on incidence of sport-related injury, Young and White suggested that the disproportionately high rates of injury among young male athletes support the notion that cultures of hegemonic masculinity promote (even demand) risk-taking behaviors, including participating in sport while experiencing significant pain and debilitating injuries. The footballer in Steve's story epitomized an athlete trying to live up to codes of dominant masculinity in sport, about which Young and White commented:

To be socialized into most dominant forms of masculinity [not only] involves learning and celebrating emotional denial, distance, and affective neutrality, but also the cultural importance of actions that often exact a physical toll. Male prowess often is based on types of physicality that are frequently destructive . . . [and] often involve conspicuous silences around health. As a result, sensitization to bodily well-being and matters of preventive health in general become viewed as the jurisdiction of women and "ambiguous" men. (p. 113)

In Steve's story, the constructed footballer talked about times when he was forced to play despite pain and was pushed into participation by humiliation and coercive tactics. Living up to cultural ideals of dominant masculinity in sport, however, is not just an issue for males. Both male and female athletes represented in the stories of John and Jane appeared to feel the pressure to tough it out when injured, sick, or excessively fatigued. All of these athletes seemed to buy into, or at least give into, the "no pain, no gain" and the more-is-better attitudes constantly trumpeted in their sports. Furthermore, athletes represented in all three tales alluded to the conspicuous silence around injuries reinforced by their coaches and parents. Although they did not talk specifically about OT, Young and White hinted at the role of dominant masculine cultures in contributing to OT-related injuries:

Overuse injuries are also examples of negative health outcomes that may be associated with dominant forms of masculinity, but not necessarily with sports that involve direct violence to the body. Some male athletes construct alternative ways of masculine identification by focusing more on endurance than aggression. (p. 116)

Aside from the footballer in Steve's story, the other male and female athletes represented in the three stories were not involved in violent contact sports, yet they still turned to overtraining behaviors, such as increasing training efforts, to assert their identities as tough athletes. Young and White (2000) stated, "There is currently a significant silence in the culture of *male* [our italics] sport about the physical toll exacted on players in the process of sport-related masculinization" (p. 123). From the athlete interviews in this book, it seems that both male and female athletes are pushed to keep silent about pain and injury and to conform to the entrenched masculinization processes inherent in the arenas of competitive sport. Another feature in the athletes' stories is that most of them had male coaches, but gender does not determine coaching behaviors so much as cultural and historical traditions do. Female coaches are just as susceptible to the masculine cultural pressures as male coaches are (Cohen, 2001; Messner, 1990).

Thus, it is important to recognize and combat the powerful contribution of sport cultural processes to injury, mismanagement of injury, and overtraining behaviors if we are to cope effectively with (and make adaptive decisions about) training and recovery, as well as to respond effectively to injury, illness, fatigue, and other setbacks.

Personal Factors

Athletes described here and in the literature appear to share similarities in beliefs, behaviors, and personality factors associated with risks for overtraining. Wrisberg and Johnson (2002) provided quotes from an overly motivated college tennis player who appeared to believe that he always had to push himself harder, disregarding physical limitations, with the result that he overtrained to the point of injury several times. Similarly, a number of athletes represented in the stories of Steve, John, and Jane could be characterized as supermotivated, holding beliefs that training harder, even in the face of pain and serious injury, is the way to succeed in their sport. In John's story, one athlete mentioned using the mantra "rest is for the dead." Wrisberg and Johnson also presented quotes from a runner who believed that taking a day off was equivalent to being weak, and in Jane's story an athlete who viewed taking extra recovery as a sign of weakness said she loathed rest days and tried to avoid complete rest days at all costs.

Numerous athletes with experiences of overtraining described themselves as perfectionists. Scanlan, Stein, and Ravizza (1991) quoted an elite figure skater: "I was a perfectionist. . . . I would never accept myself not doing it perfectly" (p. 115). Several athletes in Jane's story also talked about perfectionist tendencies and wanting to have perfect relationships. The logic of perfectionist fantasies runs like this: "If I am perfect at my sport, then I will be perfectly loved by others. All my relationships will become perfect, if I can just do this one thing [perform] perfectly."

These athletes' strivings for perfection illustrate how the pursuit of an ideal can fuel overtraining behaviors. On a deeply personal level, both the

Looking at a number of these personal characteristics and behaviors, I feel a need to offer a caution about interpreting athlete characteristics as risk factors for overtraining. As mentioned previously, the story of Jane highlighted athletes who appeared to be driven to overtraining from within, but who developed these internal drives during their formative years when trying to appease often-abusive demands, and garner the contingent love and approval, of parents and coaches. With this caution in mind, I think it could be useful to assess the personal characteristics of athletes that might make them susceptible to overtraining, while maintaining awareness of the origins of those characteristics.

The composites of Steve, John, and Jane show that athletes' personality factors, behavioral tendencies, and personal beliefs are hugely significant in understanding susceptibility to OT. Nonetheless, it is equally important to understand what created these factors, tendencies, and beliefs in the first place (e.g., the influences of coaches, parents, and sport culture).

CONCLUSIONS

The goal of this chapter and this section of the book is to put a truly human face on OT and on OT processes. We need theory and research to help us understand the multifarious factors and conditions that lead athletes down damaging paths, and many scientists in exercise physiology and sport psychology have helped immensely in gathering knowledge about OT. Sean, Mark, and Tony are all teachers. We all teach in classrooms, and we all supervise other students in their sport psychology placements. Teaching is what we do. We have found over the years (as students and later as teachers) that theory and research are central to understanding human behavior, and we would never leave home without them as guides for our teaching behaviors. Time and again, however, we have found that what makes theory and research stick in the minds of our students are the stories we tell of human lives and encounters in the extremely messy, confusing, and complicated real world of sport. We all have the experience of talking to athletes and suddenly thinking, "This story reminds me of that tale we heard years ago in class."

Humans are storytelling animals, and storytelling probably has significant survival value. Stories instruct, warn, amuse, sadden, enrage, and motivate. Stories have the potential to pass on wisdom and compassion. As for the tales in this part of the

It's common for athletes to have negative attitudes toward rest. A runner in one study believed that taking a day off was being weak. Attitudes like this can lead to exhaustion and even injury.

Krane et al. (1997) gymnast and an athlete in Jane's story talked about resorting to overtraining and self-harming, perhaps as compensatory behaviors to cope with existential angst. The gymnast said, "I purposely hurt myself to make myself better, to make myself feel like I was existing" (p. 65). In a similar vein, an athlete in Jane's story said, "I don't have many other things; so I was overtraining myself. . . . At least it will take my mind off how bad other stuff might be." These two athletes seemed to be struggling with similar issues, trying to use sport to heal their feelings of emptiness. Both, however, tried to make themselves better through self-harm and overtraining, methods that they had learned from their coaches and parents. These athletes saw themselves as people who could endure more than others, even if such attitudes and their corresponding behaviors led to the athletes' ultimate downfall.

book, we are not so sure about how much wisdom they contain, but we firmly hope they help readers develop some understanding of and compassion for athletes who become ensnared in cultural imperatives and patterns of overtraining behaviors that ensure that what they fear most will happen (failure), does happen.

IV

PAST MODELS AND CURRENT CONCEPTIONS

In this final section of the book, we pull together ideas, research, and personal stories from the previous chapters and synthesize what we have come to understand about overtraining (OT) into a comprehensive model. We discuss the two most prominent current models of overtraining (Kenttä & Hassmén, 2002; Myers & Whelan, 1999) and integrate them into our work. At first glance, the model we developed may be a bit overwhelming; with all those circles, boxes, arrows, and variables so interconnected, where does one even begin? The model is more digestible when taken apart piece by piece, and that is what we do in chapter 11.

As we often tell our students, "Any complicated human phenomenon, in sport or elsewhere, when carefully explored, becomes even more complex and convoluted." Overtraining is not an isolated phenomenon, nor does it occur in a vacuum. For example, OT cannot be reduced to a behavioral system of contingencies of reinforcement and punishment. All models of human behavior and interaction are necessarily incomplete, and model building is influenced by a variety of factors, including the zeitgeist (spirit of the time) and the personal bents of the model developers. For example, the zeitgeist of psychology in 1960s America was behaviorism.

A comprehensive model of overtraining developed by psychologists from that period and place would, most likely, have been substantially Skinnerian (after psychologist B.F. Skinner)—still complex, but housed within a behavioral framework. On the issue of personal bents of model developers, we (the authors) are all psychologists with cognitive-behavioral and psychodynamic orientations, and our model reflects those interests. If we were exercise physiologists, then the model would probably have different variables featured prominently, such as resting heart rate and other physiological markers of overtraining.

Using models to understand and explain real-world phenomena can be problematic, because models are usually clean and neat, whereas the real world is quite messy. Fitting data and experiences from real people in the real world into a model is often a procrustean act. Procrustes is a character in Greek mythology whose name means "he who stretches." He kept an inn for travelers and had a famous one-size-fits-all bed. The bed, however, did not magically adjust to each person; rather, Procrustes adjusted each traveler to fit the bed. If a guest was too short for the bed, then Procrustes would put him on a rack and stretch him to fit. If the guest was too long, then feet or legs would be cut off. Fitting the real world into models involves similar (but we hope less gruesome) processes.

In the short final chapter of the book, we consider what we have found and point to future directions, both in research and in application to professional practice. We are not interested in attempting to fit athletes' and coaches' experiences into the model. Rather, we are pragmatists who want to see how *useful* the model is when we encounter athletes who may be at risk, or when we operate in environments and cultures that increase vulnerability to overtraining. We like to think of the model as a fluid tool for understanding, and we see this tool as subject to continual modification. Thus, chapter 11 may appeal more to researchers than to lay sports people, but chapter 12 should have a broader audience of individuals who are concerned with sport, health, performance, happiness, and the well-being of athletes.

11

MODELS OF OVERTRAINING
Then and Now

In writing this chapter, we felt exceedingly challenged by the task of bringing together the results of the expert and athlete interviews, and of presenting a clear picture of risk factors for overtraining (OT). We have gained an appreciation for the work of other researchers and authors who have attempted to depict the complexity of OT in conceptual models (i.e., Kenttä & Hassmén, 1998, 2002; Meyers & Whelan, 1998), and we hope that we may add to the literature by expanding on the work of these researchers and by presenting a descriptive model of OT risk factors, processes, and outcomes. In the following sections, we present a synthesis, uniting, as much as possible, the perspectives of the experts with the experiences of the athletes. We then present a model of OT risk factors, processes, and outcomes based on the expert and athlete interviews, and we critically reflect on the OT conceptual models of Meyers and Whelan and of Kenttä and Hassmén.

SYNTHESIS OF EXPERTS' PERSPECTIVES AND ATHLETES' EXPERIENCES

In general, the expert interviews provided a real-life background for understanding the broad range of potential OT risk factors, whereas the athlete interviews provided in-depth illustrations of a smaller number of those risk factors affecting individual athletes. The experts provided a comprehensive list of personal and situational risk factors, and the athletes described the personal, and sometimes painful, experiences of overtraining. The athletes' stories put a human face on the risk factors outlined by the experts, providing glimpses of the dynamic, complex interactions among people, situations, and sociocultural factors that influence overtraining behaviors and outcomes. The athletes' stories also provided a sense of the omnipresence of overtraining pressures and illustrated both the obvious and the not so obvious influences that constantly drive athletes' behaviors.

Both experts and athletes emphasized the roles of coaches and parents in instigating, teaching, reinforcing, and maintaining OT behaviors and highlighted the position of injury as both a cause and a consequence of OT processes. Furthermore, both experts and athletes illustrated how coaches and parents could be key drivers of maladaptive behaviors surrounding injuries. With respect to the significance of injury and illness issues in overtraining, O'Toole (1998) has commented on the misattribution of causes of injury and illness in sport:

> Increased susceptibility to musculoskeletal injury or infections, such as head colds, may be indicators of a state of overreaching or overtraining, but may be misinterpreted as isolated, local problems rather than manifestations of the overtraining syndrome. (p. 13)

This statement seems even more relevant to us now, in light of the experts' and athletes' emphasis in this book on injury in OT, as well as their comments about illness. Perhaps most researchers would acknowledge that injury and illness are significant consequences of overtraining, but our research suggests that injury might be the most common outcome of overtraining processes. This connection of injury with OT seems to be a promising direction for more systematic study in the future.

The interaction between the dynamic influences of parents and coaches (e.g., parental aspirations for success through their children, coaches' use of children for career advancement) and the OT outcomes of injury and illness also highlights the significance of interpersonal contributions to athletes' responses to potential OT situations. In discussing markers of OT, the experts pointed out that often athletes who experience the outcomes of OT will become emotionally distressed and reactive, which leads them to poor decision making in response to unpleasant outcomes and thus may prompt continued OT behaviors. One expert referred to such maladaptive coping by athletes as the "need-to-please disease." The athletes' stories also illustrated how athletes' anxieties about what parents, coaches, and others think of them might lead them to respond, desperately, in maladaptive ways to injuries, illnesses, and other setbacks.

Despite obvious overlaps in the experts' and athletes' descriptions about parents and coaches and about issues surrounding injuries, we noted differences in the level of detail emerging from the two groups. Experts mostly gave a sense of OT risk factors to look for in athletes in the *here and now*, whereas the athletes' stories emphasized their unique familial and psychological histories. The characteristics and experiences of the athletes making up the Steve, John, and Jane tales, as well as Sean himself, shed light on the depth and breadth of pathogenic processes in athlete behaviors that are instigated, promoted, and maintained by parents and coaches. The athletes whose stories were the bases for the three composite tales also exemplified how individuals may turn to competitive sport in attempts to escape their own feelings of inadequacy and emptiness, look to success at major sporting events to elevate their self-esteem, and be driven by cultural and subcultural forces dominant in sport, society, the media, and their relationships with parents, coaches, and teammates.

The integration of the expert and athlete findings could enhance understanding of past and present experiences with OT processes and outcomes and help predict possible future responses to OT situations. We hope that this project might also help answer our original questions about why athletes overtrain in the first place and continue to do so even in the face of balanced guidance. In conducting OT risk assessment, one might gather information on what happened in an athlete's past that led to the development of OT behaviors, then ask what is going on in the present to drive or maintain such behaviors. In particular, one could look at circumstances that elevate stress (requiring increased recovery) or motivate increases in training (also requiring increased recovery), and at the influences that reinforce or maintain maladaptive responses to stress–recovery imbalances, injury, illness, or other setbacks. Finally, one could ask: How might we anticipate future responses to OT situations?

After completing the interviews, we initially had the sense that OT could be broken into two major categories: personal and situational risk factors. The personal risk factors would involve what athletes bring to the table, and the situational risk factors would involve the aspects of their environments that push athletes to overtrain. Nevertheless, with a shift away from identifying physical training as the major stressor that leads to overtraining, and toward identifying a whole range of training and nontraining stressors—as well as factors associated with underrecovery, as emphasized by Kellmann (2002), Kenttä and Hassmén (2002), and others (e.g., Lehmann, Baur, Netzer, & Gastmann, 1997)—it became apparent that these two categories (i.e., personal and situational) were not specific enough, as overarching dimensions, to depict OT risk factors. Thus, trying to focus on the central issues emerging from the project, we started to conceptualize OT risk factors in terms of intra- and interpersonal, situational, and sociocultural influences, both past and present, on athletes' behaviors. We could also see that all of these influences interacted in ways that could either motivate or push athletes to increase training, or create circumstances that demanded increased recovery, as indicated in the general dimensions of the experts' OT risk factor list. In the next section, we attempt to depict these OT risk factors, the early signs that represent stress–recovery imbalances, the behavioral responses to those signs and symptoms, and the potential outcomes in a dynamic OT model.

THE OT RISKS AND OUTCOMES MODEL

In trying to depict a model of OT risk factors, early signs, behavioral responses, and outcomes, we struggled with how to illustrate the various aspects of the model schematically and efficiently, while preserving the complex, dynamic, and interactive qualities. We drafted numerous network diagrams and flow charts by hand, shifting from the convoluted and confusing to the brief and simplistic. At one point, we thought we had settled on a temporal model describing risk factors within the framework of an athletic season, but decided that the temporal element could be limiting. Consequently, we have worked out a general descriptive model, which we refer to as the OT Risks and Outcomes Model, and have followed it up with a description of how the model might be applied, temporally, across an athletic season.

Background of the Model

We were cognizant of integrating the experts' and athletes' stories and, at the same time, were mindful of the existing literature on OT. As mentioned in chapter 2, researchers have focused on trying to identify markers of OT syndrome (e.g., Hooper & McKinnon, 1995; Uusitalo, Huttunen, Hanin, Uusitalo, & Rusko, 1998), potentially limiting themselves to looking at the immediate states of athletes' fatigue. These limitations might not have allowed investigators to develop a broad picture that includes what athletes bring to any training situation from their personal histories. The focus on immediate fatigue states also might have limited researchers to questions that did not involve how athletes could be expected to respond during times of increased stress; how athletes normally cope with injury, fatigue, or illness; or what activities athletes engage in during their own time outside of the coaches' programs. Answering the question of what drives athletes to overtrain seems to require in-depth life histories, which reveal attitudes and behaviors toward training and recovery learned during formative years in sport. Gathering such detailed information could shed light on past experiences with, and responses to, injury, illness, and fatigue. Furthermore, conducting biographical analyses of athletes could illuminate common behaviors surrounding OT, perceived external pressures to engage in OT behaviors, and struggles with unfulfilled needs that athletes are trying to satisfy by participating in competitive sport.

In bringing together the athletes' and experts' stories, it became apparent that assessment of OT risk would be an ongoing process, influenced by dynamic interactions of past experiences, athlete characteristics, and situational pressures. Furthermore, these interactions take place within a socio-cultural context that underlies and influences decision making and behavior with respect to OT.

As we developed the OT Risks and Outcomes Model, we referred to the OT risk factor list from the expert interviews to provide an outline for exploring past and present personal characteristics and situational variables that put athletes at risk for overtraining. From the athlete interviews, we were guided to represent athletes' past experiences with coaches and parents in the model and to depict the influences of such others on athletes' responses to current training and recovery situations. We were also guided to include illustrations of the pressures inherent in sport cultures.

The OT Risks and Outcomes Model is illustrated in Figure 11.1. We have divided it into four parts:

- Risk factors
- Early signs
- Behavioral responses
- Outcomes

We discuss each of the parts of the model in the next four sections.

Risk Factors

The first part of the model shows the major categories of OT risk factors in dynamic interactive relationships that influence athlete beliefs and behaviors, leading to the early OT signs depicted in the second part of the model. We had originally divided OT risk factors into athlete characteristics, situations that pressure athletes to increase training, and situations that affect athletes' needs for recovery. In synthesizing the experts' opinions with the athletes' stories, however, it seemed that these original divisions did not capture the way the dynamic interactions between different systems influenced athlete behaviors. To assess the OT risk of any particular athlete, it seemed that we could start by looking at the athlete's characteristics, behaviors, and beliefs. To gain further insight into why the athlete behaved in certain ways, or held particular beliefs, with respect to training and recovery, we could then look at how (past and present) interpersonal and situational factors were influencing the athlete's behavior. Finally, in completing risk assessment, it appeared important to acknowledge the potentially strong influences of the sociocultural context (the culture of the sport) in which the athlete behaviors, interpersonal influences, and situational factors were embedded. In trying to represent a comprehensive picture of the many different factors and interactions affecting

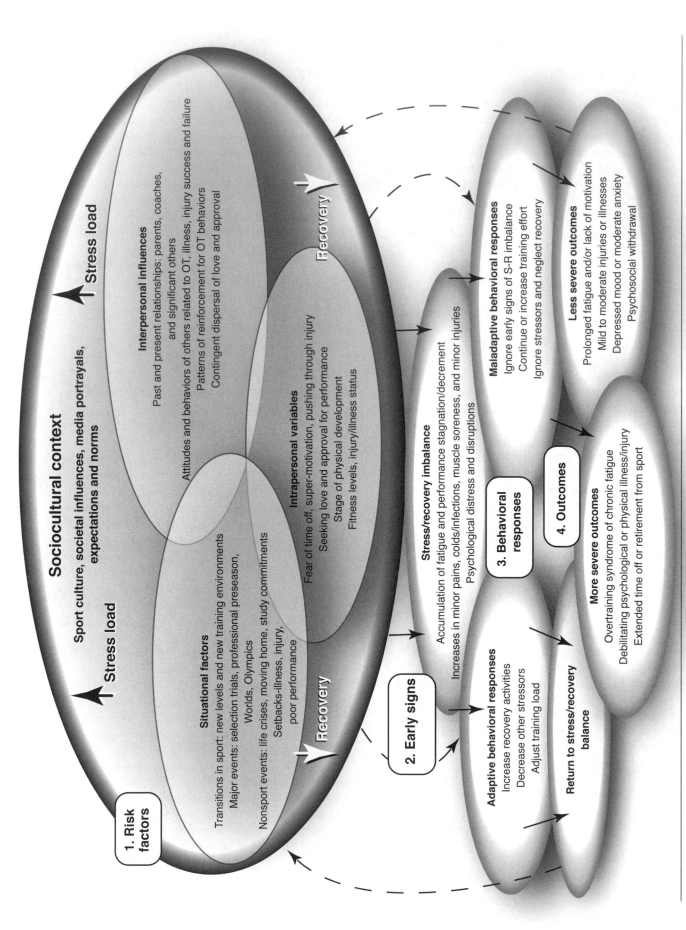

FIGURE 11.1 OT Risks and Outcomes Model.

OT processes, then, we labeled the major categories of risk as follows:

- Athlete intrapersonal variables (e.g., personality)
- Interpersonal influences (from coaches, parents, and significant others)
- Situational factors (e.g., upcoming competitions, team selection)
- Sport sociocultural context or environment in which the other factors are imbedded

It also seemed important to show that the different categories of influence could either motivate an athlete to increase training or increase demands for recovery, either of which might create circumstances that could upset an athlete's stress–recovery balance (illustrated in the model by the changes in shading of the risk factor circle and the labels for increased training stress load and decreased recovery). Furthermore, for any athlete at any given time, one or more of these categories could have variable and disproportionate influences on OT processes. Unfortunately, such temporal variations could not be shown in a static, two-dimensional diagram. For example, in the athletes' stories, despite overlap among themes, there were examples of different primary drivers of OT: Steve was driven by the tough culture of professional sport, John was spurred on by upcoming Olympic Games, and Jane was pushed excessively by parents and coaches.

Early Signs

The second part of the model represents the early signs and processes that evolve from interactions between intrapersonal, interpersonal, situational, and sociocultural influences, and which lead to initial imbalances between stressors and recovery. According to the experts, monitoring these initial signs and phases of stress–recovery imbalance could include attending to a number of different markers when assessing athletes' current physical and psychological states. They suggested evaluating current *physical* states by looking at increasing levels of fatigue; unexpected decreases in performance; evidence of minor niggles, pains, muscle soreness, colds, and infections; changes in physiological markers associated with stress (e.g., resting heart rate); changes in routine; sudden weight loss or weight gain; and counterproductive biomechanical or technique changes. The experts suggested evaluating athletes' current *psychological* states by observing them for emotional distress

or reactivity; fears of failure; guilt about missed or reduced training; and anxiety related to communicating with coaches or others about fatigue, injury, illness, or other stressors. Experts emphasized the importance of using OT markers not as conclusive evidence of OT, but as starting points for investigating the possibility that OT processes are occurring.

Behavioral Responses

The third part of the model represents how athletes respond to the initial experiences of early OT signs and stress–recovery imbalances. The arrows on the model from section 1 (the sociocultural context oval), combined with the arrows from section 2 (stress–recovery oval) and feeding into section 3 (behavioral response ovals), are meant to show that athletes' behavioral responses to OT processes and stress–recovery imbalances are influenced by psychosocial factors from section 1. In other words, how athletes react to initial signs of excessive stress or fatigue is determined by what they believe, what they have experienced, how they are being influenced by others, what situations might limit or motivate them, what signs and symptoms they develop, and how they are predisposed by the sporting culture. It is conceivable that some athletes might have quite balanced approaches to training and recovery, and that they might also have people around them with similarly adaptive approaches. Nonetheless, when major events—such as national selection trials, World Championships, or Olympic Games—loom in the near future, such balanced athletes could be driven to push harder than they did previously, with the result that they engage in maladaptive behavioral responses to stress–recovery imbalances and consequently experience adverse outcomes. In other cases, athletes could be driven to maladaptive behavioral responses because of abusive coaches or parents, or because of the constant barrage of cultural imperatives present in the sporting world, in the media, and in other forms of communication.

Outcomes

The fourth section of the model represents potential outcomes of athletes' responses to stress–recovery imbalances. In the event of adaptive responses to initial stress–recovery imbalances (see the adaptive behavioral response oval), athletes would conceivably return to balance between their stressors and recovery. The dotted arrow leading back to section 1 suggests that the experience of an adaptive outcome could influence future interactions of risk factors. In the event of maladaptive behavioral responses to initial stress–recovery imbalances, which are influenced by the psychosocial factors from section 1, athletes would likely spiral downward to more serious adverse outcomes, including substantial performance decrements, injuries, illnesses, psychological distress (e.g., anxiety, depression), and fatigue syndromes (described as OT syndrome or chronic fatigue). We have divided the outcomes of maladaptive responses into less and more severe outcomes. The idea is to show that, with some injuries, illnesses, instances of minor psychological distress (e.g. depressed mood, mild anxiety, social withdrawal), and levels of fatigue, athletes might still continue to train, albeit with difficulty. These adverse outcomes could then become new situational risk factors influencing future OT behaviors and beliefs, depicted by the dotted arrow looping back to section 1. These setbacks could also be seen to drive further OT processes, thus adding to vicious cycles of negative influences, more OT signs, maladaptive responses, and adverse outcomes. With more serious injury, illness, psychological distress (e.g. full-blown depression, debilitating anxiety disorders, social isolation), and fatigue syndromes—as depicted in the final level of maladaptive outcomes—athletes could need to take complete breaks from training, often for prolonged periods of time, and in some cases might be forced into retirement from competitive sport. In the worst case, with severe depression from a loss of identity and meaning in their lives, athletes could turn to suicide as a final and tragic outcome.

Describing OT Risk Assessment within a Temporal Framework

In the following sections, we attempt to describe an application of the OT Risks and Outcomes Model for OT risk assessment using a temporal framework. A yearlong approach to OT risk assessment, monitoring, and intervention could involve working to understand athletes' beliefs and behaviors surrounding overtraining, as well as the interpersonal, situational, and sport culture influences on those beliefs and behaviors. One could look at OT risk assessment as an ongoing process throughout athletes' training cycles. For this example, we have tried to describe OT risk assessment for individual athletes in different scenarios across a competitive year, starting with the commencement of training after an off-season break. In conceptualizing risk assessment for OT, we hope to simplify the ongoing

dynamic interplay of influences by dividing risk assessment into four possible scenarios during an athlete's year:

- Planning training and recovery strategies at the start of a new season
- Monitoring stressors and recovery activities and adaptations throughout the training cycle
- Responding to the first signs of stress–recovery imbalance
- Responding to setbacks, such as injury, illness, psychological distress, and excessive fatigue states, if they occur

Planning Training and Recovery

At the beginning of a new season or training cycle, athletes could be assessed in terms of what they bring to the table. Questions posed to athletes at this point could be focused on

- their attitudes toward training and recovery;
- their past interactions and relationships with coaches, parents, and significant others;
- their history with injury, illness, and fatigue;
- their past training programs and responses to them;
- their expectations about progression in the sport;
- their perceptions of others' expectations of them;
- their current levels of fitness, including any chronic injuries or illnesses; and
- their current states of physical development (e.g., young athletes going through growth changes, or older athletes dealing with reduced recovery capacities).

The idea is to get a picture of what athletes can tolerate in terms of training load and how they might be expected to respond to various outcomes of training and recovery throughout the year. For example, in the event of new injuries, having knowledge about athletes' past responses to injuries, which might have involved anxious attempts to return too early to training, could help coaches, doctors, or sport psychologists guide athletes in effective rehabilitation processes and encourage timely, rather than premature, returns to sport.

Initial risk assessments at this time of the season—based on the intrapersonal, interpersonal, and situational risk factors presented in part 1 of the model—could set the stage for monitoring of, and early intervention in, athletes' overtraining behaviors as the season progresses. Coaches and sport psychology practitioners could refer to the OT risk factor list from chapter 3 as an outline to guide exploration of an athlete's past, and this list could be augmented with the themes emphasized in the athletes' experiences portrayed here in the composite case studies. The athletes' stories, especially those depicted in Jane's tale, highlight the importance of assessing past experiences with sport, as well as the perceived roles of coaches and parents in those experiences, when looking at risk for overtraining. The stories illustrate how, despite coming under the new leadership of balanced coaches who encourage adequate recovery, athletes might continue to overtrain because of maladaptive coping mechanisms learned in the past for dealing with demanding parents and coaches. In these situations, with knowledge of athletes' beliefs about overtraining and recovery, new coaches might be better equipped to develop positive communication with athletes that helps them overcome some of their overtraining patterns.

In our work with experts and athletes, we observed that athletes sometimes held information back. They undertook extra training without telling their coaches or carried niggling injuries that they did not report. The risks will always exist that an athlete is not completely honest with his or her coaches. Sometimes athletes might think that "secret" training will help them look stronger at "official" training, thus impressing the coach, or that reporting a niggling injury might lead the coach to stop them from training or even decide not to select them for an upcoming game or event. The best way we know to address the risk of athletes withholding information, in terms of the planning of training, is through education of athletes and coaches about the risks associated with overtraining. Educated coaches should be more open and understanding, emphasizing the balance between training and recovery, rather than the more-is-better philosophy. Informed athletes should appreciate the importance of telling their coaches and expert support staff about all their training-related concerns to ensure that all involved agree on an appropriate balance between training and recovery. Furthermore, athletes should have confidence that educated coaches and support staff aim to achieve that training–recovery balance, which is in their athletes' best interests.

Monitoring Stress and Recovery

From the athletes' stories, it was evident that risk for overtraining was an ongoing issue throughout the season. Thus, it could be important to carry out risk assessments periodically as a training cycle continues. This phase of monitoring could involve vigilance directed at situational variables that increase risk for overtraining—both those that motivate athletes to push harder and those that affect athletes' recovery needs. Any time that decisions have to be made about changes in training and recovery activities, it could be important to consider how athletes might normally have responded in the past, which influences how they are likely to respond in the present. Such considerations might include anticipating high-risk situations and preparing to make adaptive decisions when they arise. This phase is about risk assessment when novel situations or events arise that might affect OT behaviors.

Responding to Stress–Recovery Imbalances

In listening to the athletes' stories, there appeared to be sensitive periods during training when things started to become slightly imbalanced. Athletes' responses to these initial stages of imbalance often set them up to move to more serious stress–recovery imbalances. Thus this phase is about assessing how athletes will respond when stressors first start to exceed recovery capacities. When athletes start to build up levels of fatigue; experience small decrements in performance; experience mood and behavioral changes; and develop minor niggles, infections, or other symptoms of stress–recovery imbalances during the training cycle—it is then that we could refer to information gathered at the start of the season about intrapersonal, interpersonal, and situational OT risk factors. Knowing how athletes think and feel about themselves during times of duress could be helpful, because these early stages of stress–recovery imbalance are critical times for coaches and athletes to make sound decisions about training and recovery. If coaches and athletes respond with anxiety and desperation, hoping that hard work will overcome any problems, athletes may spiral downward to more extreme overtraining, injury, psychological problems, and other adverse outcomes.

Responding to Setbacks

Common themes in the athletes' stories included the following responses to being faced with significant setbacks, such as illness, injury, or excessive fatigue: becoming anxious and attempting to block out possible consequences of setbacks; ignoring signals in their bodies telling them to recover more; and pushing for early returns to full training (and often training and competing with serious pain or debilitation). It could be worth considering reassessment and monitoring of athletes' OT risk right at the time a setback occurs. Such assessments might help coaches and athletes minimize the chance of repeating the behaviors that resulted in the initial setbacks, or help them make the most effective decisions about rehabilitating and returning to full training. Such risk assessments could also provide more information for sport psychologists as they work to address the psychological consequences of OT within a psychotherapeutic context.

Conclusions About the OT Risks and Outcomes Model

Looking at the OT Risks and Outcomes Model, and four possible time frames during which one could assess risk throughout an athletic season, it appears that OT risk could be seen as an ongoing issue in athletes' lives. Throughout their careers, athletes' are continually influenced by the interactions of their current experiences with attitudes and behaviors formed from past experiences. These interactions form new patterns of behaviors, as well as reinforce old ones, both adaptive and maladaptive.

OT RISKS AND OUTCOMES MODEL COMPARED WITH OTHER MODELS

In the following sections, we provide outlines of the Meyers and Whelan (1998) and Kenttä and Hassmén (1998, 2002) models of OT, then compare them with our OT Risks and Outcomes Model. The former two models are stress–adaptation models, and both are aimed at conceptualizing processes that lead to OT syndrome.

Multisystemic Model

Meyers and Whelan (1998) described a systemic stress–adaptation model of OT that illustrates the importance of understanding multiple sources of stress in athletes' lives, effects of athletes' experiences in one situation on experiences in another context, and influences of interrelationships between athletes, other people, and various environments. Meyers and Whelan discussed their model in terms of "building a framework for understanding why competitors with similar physical skills and capabilities, exposed to almost identical training regimens, may demonstrate

widely variable outcomes" (p. 336). Meyers and Whelan held that understanding the OT stress experience requires us to consider the functioning of individual athletes in complex systems. They suggested that it is necessary to consider many of the environmental contexts in which athletes operate, including training and competition, as well as with social, familial, and cultural histories, and how these factors interact to influence athletes and their behaviors.

Figures 11.2 and 11.3 from Meyers and Whelan depict several of the systems surrounding athletes in sport and nonsport situations respectively. Meyers and Whelan pointed out that challenges or threats might originate within any situation or system and produce stress that can influence behavior in any other system or situation. They stated, "An athlete's failure to perform may be due to a myriad of interacting influences. Consequently, understanding the athlete's performance requires consideration of both sport and extra-sport contexts" (p. 349). Understanding the overall impact of a stressor within the multiple contexts is central to understanding any potential OT issues presented by an athlete.

Although Meyers and Whelan discussed the impact of stressors within present systems and contexts, they also noted the significance of past events and experiences, suggesting that it is important to recognize the life experiences and personal histories that athletes bring to any training situation. They also commented on the cultural context in which systems are embedded, suggesting that, within the context of overtraining, "society, culture, politics, and economics all have some effect on the athlete" (p. 350).

In general, Meyers and Whelan presented a descriptive OT model aimed at increasing awareness of the multiple sport and nonsport contexts in athletes' lives that influence OT processes and bring about OT outcomes. The model depicts the overarching systems or contexts that are important to consider in understanding OT, including athletes' past histories, present sources of stressors, interrelationships between current sport and nonsport contexts, and interactions with other people.

Conceptual Model of Overtraining

Kenttä and Hassmén's (1998, 2002) model of underrecovery and OT, depicted in figure 11.4, captures factors and processes similar to the Meyers and Whelan model by depicting interactive, multisystemic influences on OT processes and OT syn-

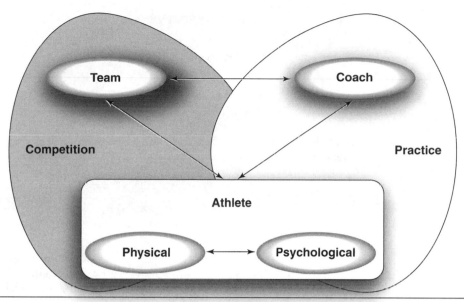

FIGURE 11.2 Systemic view of the athlete in sport contexts.

Adapted, by permission, from A.W. Meyers and J.P. Whelan, 1998. A systemic model for understanding psychological influences in overtraining. In *Overtraining in sport*, edited by R.B. Kreider, A.C. Fry and M.L. O'Toole. (Champaign, IL: Human Kinetics), 351.

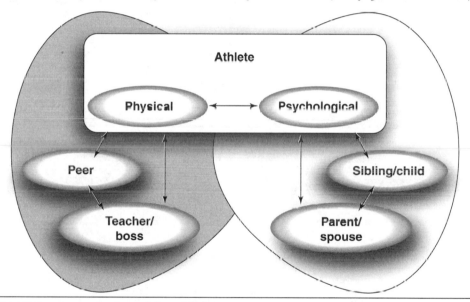

FIGURE 11.3 Systemic view of the athlete in nonsport contexts.

Adapted, by permission, from A.W. Meyers and J.P. Whelan, 1998. A systemic model for understanding psychological influences in overtraining. In *Overtraining in sport*, edited by R.B. Kreider, A.C. Fry and M.L. O'Toole. (Champaign, IL: Human Kinetics), 353.

drome. Kenttä and Hassmén, however, proposed a more practical model than Meyers and Whelan's, one they designed to be used in assessing athletes' current states of adaptation to stressors. Similar to Meyers and Whelan's project, Kenttä and Hassmén aimed specifically at conceptualizing the factors that lead to OT syndrome, which they labeled the *staleness syndrome*. To describe OT, Kenttä and Hassmén referred to the definition by Lehmann, Foster, and Keul (1993) of overtraining syndrome as "an imbalance between training and recovery, exercise and exercise capacity, stress and stress tolerance.

Stress is [considered] the sum of training and nontraining stress factors" (p. 854).

Kenttä and Hassmén defined OT processes in their model in terms of interactions between three major subsystems: psychological, physiological, and social. They suggested that within each of these systems, three components interact to affect the overall balance between stress and recovery: stress, stress capacity, and recovery. Kenttä and Hassmén gave examples of how the components can be broken into subcomponents. For example, neuromuscular capacity is described by aerobic and anaerobic energy

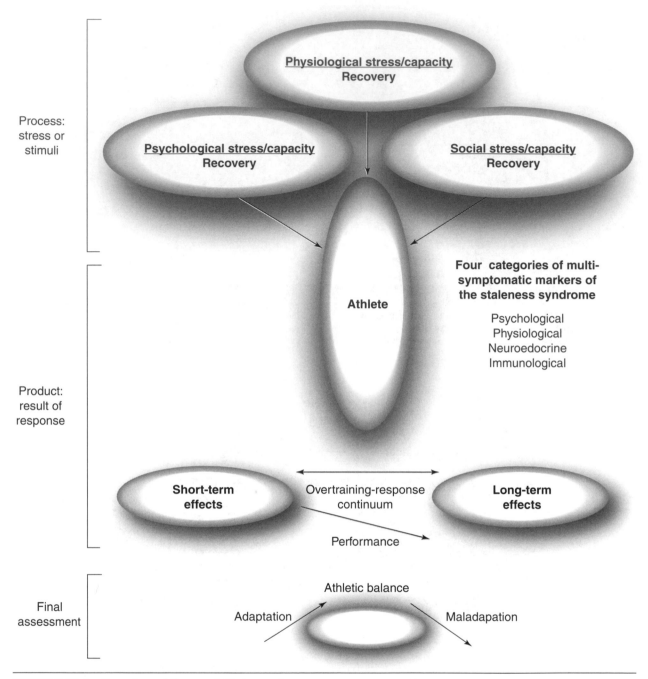

Process:
stress or
stimuli

Product:
result of
response

Final
assessment

FIGURE 11.4   Overview of the overtraining and recovery process.

Adapted, by permission, from G. Kenttä and P. Hassmén, 2002, Underrecovery and overtraining: A conceptual model. In *Enhancing recovery: Preventing underperformance in athletes,* edited by M. Kellmann (Champaign, IL: Human Kinetics), 68. Based on Kenttä and Hassmén, 1998.

production, general strength, specific strength, and technique. Capacity for psychological stress is described by level of self-confidence, capacity to cope with anxiety, attentional capacity, motivational level, attitude control, positive mental health, and visualization skills. The capacity to handle social stress is described by ability to create, negotiate, and maintain relationships with others.

In Kenttä and Hassmén's model, the interaction of the many processes leads to responses and

outcomes that can be assessed along a continuum moving from adaptation to maladaptation. Looking at this continuum, Kenttä and Hassmén (1998) summarized the specific aim of their model in monitoring OT and recovery processes:

The aim of monitoring training and adequate recovery in the elite athlete is to reach a balance in the zone where training yields optimal increases in performance.

. . . We suggest that this optimal zone be defined as the *adaptation threshold*. Theoretically, the adaptation threshold is the dynamic 'breaking point' where adaptation suddenly becomes maladaptation. Together, recovery, stress, and capacity can be viewed as three variables affecting the adaptation threshold. Thus, the need to identify the individual's dynamic threshold can be seen as the overall goal in monitoring training and recovery. (p. 13)

Kenttä and Hassmén suggested that the way to remain in the optimal zone of adaptation, or to increase the adaptation threshold, is for athletes to optimize recovery activities and improve stress tolerance (they suggested that reductions in training loads for elite athletes are generally not feasible). They outlined three approaches to optimizing recovery:

- Matching recovery activities with the specific type of stressor
- Improving specific capacities, such as coping skills, to improve stress tolerance
- Minimizing psychological and social stressors

They also suggested monitoring symptoms and markers of overtraining to aid in making decisions about alterations in the training program.

Kenttä and Hassmén's (1998, 2002) model appears to be focused on assessing OT syndrome *right now*. The model provides a conceptual, holistic approach to determining an athlete's present status and current levels of stressors, capacities, and recovery activities. The model is about providing answers to the following questions: How can one identify whether an athlete is becoming overtrained, or developing a stress–recovery imbalance? Is it okay for an athlete to continue his or her training schedule? The answers to these questions are found by monitoring and evaluating the three OT processes (physiological, psychological, and social stressors, recoveries, and capacities) and associated symptomatic markers.

OT Risks and Outcomes Model Versus Multisystemic and Conceptual Models

Meyers and Whelan (1998) originally laid the foundation for considering multiple contexts influencing OT syndrome by describing interactions between general variables in both sport and non-sport contexts. Kenttä and Hassmén (1998, 2002) offered a more focused, practical assessment model illustrating how physiological, psychological, and social systems and subsystems might interact to upset athletes' stress–recovery balances and result in OT syndrome. With the OT Risks and Outcomes Model, based on the findings of our research, we have attempted to broaden perspectives on OT processes by including risk factors, early signs, adaptive and maladaptive behavioral responses, and multiple OT outcomes in one model. We have tried to illustrate that OT processes lead not only to fatigue syndromes, such as OT syndrome, but also to illness, injury, and negative psychosocial consequences. We have also attempted to incorporate themes and categories derived from our expert and athlete interviews into the components of the risk model, providing specific examples of variables that make up OT risk factors, early signs, behavioral responses, and outcomes. With the OT Risks and Outcomes Model, then, we hope to augment previous research on identifying OT processes and outcomes with an increased understanding of *why* athletes engage in overtraining behaviors, and what might influence them to continue engaging in those behaviors even when they receive support and encouragement to balance their stressors and recoveries.

The most substantial part of the OT Risks and Outcomes Model—that is, the risk factors—shares similarities with both the Meyers and Whelan and the Kenttä and Hassmén models. The interactions we describe between athlete intrapersonal variables, interpersonal influences, situational variables, and sociocultural influences parallel the structure underlying Meyers and Whelan's and Kenttä and Hassmén's perspectives, in which they described OT processes as functions of the interactions between multiple systems and contexts. Nonetheless, the novel contribution of the OT Risks and Outcomes Model is the emphasis placed on *past* formative experiences that drive OT behaviors and processes, especially the dynamic influences of parents, coaches, and significant others. Furthermore, the model is aimed at providing specific (albeit nonexhaustive) examples of variables that have been shown in the research to influence OT processes and outcomes. In a sense, the results of our project provide specific detail about the numerous possibilities of stressors and influences that fit within the overarching contexts of Meyers and Whelan's model and within the psychological, physiological, and psychosocial subcomponents of stress and recovery in Kenttä and Hassmén's model.

Although Meyers and Whelan discussed the influences of coaches and parents, and suggested deeper psychodynamic drivers of overtraining behavior, their model leaves it up to the reader to determine how and why these drivers lead to overtraining. We think that Meyers and Whelan presented a model designed to stimulate more research into the multisystemic context of OT syndrome, and perhaps push future researchers to search for answers to these *how* and *why* questions. Their conceptual model helped us develop this project, and we were motivated to find answers to their questions regarding what distinguishes one athlete from another in susceptibility to OT syndrome.

The early signs part of the OT Risks and Outcomes Model is reflective of Kenttä and Hassmén's model, in that it represents the current level of adaptation occurring in an athlete. As in Kenttä and Hassmén's model, detecting the early OT signs involves monitoring of physical and psychological variables in athletes. The added component of our model is emphasis on the constant influence of the risk factors that underlie the development of these early OT signs. That is not to say that Kenttä and Hassmén did not emphasize the influence of multiple interacting variables, because they did discuss such complexities. Rather, Kenttä and Hassmén appeared to emphasize the athletes' present states, without much reference to past experiences and dynamic influences on motivation and decision making. The current model is about expanding the view of drivers and causes of OT—from looking at present stressors, signs and symptoms, and recovery activities to looking at many past and present motivators, stressors, and influences. The model also presents the idea that, at any time, any of several categories of factors—intrapersonal, interpersonal, situational, or cultural—might be the major influence on the many different OT processes. From considering the *capacity* aspect of Kenttä and Hassmén's description of stress–recovery balance, it could be inferred that this part is associated with past experiences and behaviors, but Kenttä and Hassmén did not emphasize athletes' pasts. Rather, using the methods of monitoring training exertion and recovery, they emphasized current stressors and current recovery, perhaps assuming that capacities are more or less stable, and that they can be either risks or resources.

The capacity aspect does provide opportunities for discussing injuries, illnesses, and psychological stressors as antecedents of OT. Conceivably, the physiological capacity of an athlete is reduced during times of illness and injury, and social and psychological capacity might also allow for discussions about influences of past experiences on choices of coping mechanisms. Nonetheless, Kenttä and Hassmén did not go into great detail on these issues.

In the early signs and the outcomes sections of the OT Risks and Outcomes Model, we have included injuries and illnesses as causes of stress–recovery imbalances, as potential markers of OT, and as significant outcomes of OT processes. We understand that previous models of OT have been directed at conceptualizing and identifying processes that lead exclusively to OT syndrome, or what has also been called unexplained underperformance syndrome (Budgett, 1998), underrecovery (Kellmann, 2002; Kenttä & Hassmén, 2002), and the staleness syndrome (Kenttä & Hassmén, 2002), and the originators of such models might have purposely left out discussions of injury or illness because they were interested only in the outcome of OT syndrome. Kenttä and Hassmén do discuss illness and immunological changes in OT, but those processes and outcomes are not as central to their model as they are to ours. Given the frequency of injuries and illnesses in competitive sport and their relations to the same processes that lead to OT syndrome, as illustrated in the experts' and athletes' stories, it seems important to include them in any discussion of OT.

In addition to illnesses and injuries, we have included psychosocial issues in the early signs and the outcomes parts of the model, as markers and consequences, and have tried to depict psychosocial issues as influential variables in the behavioral responses part of the model. Although Kenttä and Hassmén (1998, 2002), Meyers and Whelan (1998), and many other researchers (e.g., Berglund & Säfström, 1994; Fry, Grove, Morton, Zeroni, Gaudieri, & Keast, 1994; Hooper, Mackinnon, & Hanrahan, 1997; Hooper, Mackinnon, Howard, Gordon, & Bachmann, 1995; Morgan, Costill, Flynn, Raglin, & O'Connor, 1988) have highlighted psychosocial issues as playing important roles in OT, much of the discussion has revolved around psychological distress, usually measured by a mood inventory such as the POMS, as a marker of impending OT syndrome. Kenttä and Hassmén enhanced this discussion by pointing out the contribution of psychosocial stressors in upsetting stress–recovery balances and leading to OT syndrome, and Hooper and Mackinnon (1995), O'Toole (1998), and Steinacker and Lehmann (2002) provided OT definitions that included depression and general disruptions of psychological well-being in their descriptions of possible adverse outcomes associated with OT. Nevertheless, there does not appear to be an OT model or framework that shows the psychosocial issues concurrently as causes, markers, and consequences of OT. We hope that by including the multiple functions and positions of psychosocial issues in our model, we might improve understanding of why and how athletes begin and continue to overtrain.

The behavioral responses section of the OT Risks and Outcomes Model may be seen as augmenting Kenttä and Hassmén's work on the underlying factors of OT processes, because it highlights the influence of psychosocial components (from part 1 of the model) on behavioral responses to stress–recovery imbalances, and prompts exploration into why and how athletes might respond in the ways they do when they experience such imbalances. Answering these why and how questions could be an integral part of interventions targeted at minimizing OT processes and outcomes. Looking at *how* the multiple systems and contexts in athletes' lives interact and operate, and *why* athletes continue to overtrain despite knowledge and advice to the contrary, could help people observing and working with the athletes—including coaches, parents, sport administrators, and sport scientists—to act effectively in guiding athletes to more balanced lives. For example, when athletes begin to experience stress–recovery imbalances, it is not the imbalances that prompt them to push harder or neglect recovery; more likely, continued OT behaviors are prompted by the anxiety they feel about being unable to live up to their own expectations or about disappointing their parents and coaches.

In some ways, we think of our project, and the emerging OT Risks and Outcomes Model, as synthesizing many different issues in OT and as representing a series of case studies that illustrate combinations of the Meyers and Whelan and the Kenttä and Hassmén models. The experts interviewed here presented quite a list of systems or contexts that influence OT behaviors and within which athletes operate. The athletes' stories, in turn, showed the idiosyncratic flavor of individuals overtraining within those systems and contexts. Our model could augment Kenttä and Hassmén's model with respect to monitoring for OT syndrome and other OT outcomes. We hope that the current model might stimulate the monitoring of OT-related injuries, illnesses, and other negative psychological outcomes (e.g., if an athlete experiences muscle tightness, is assessed to have a stress–recovery imbalance, and displays anxiety about the situation, then coaches and others might be alerted to intervene with increased recovery and prophylactic treatment). The model might also help in understanding and anticipating OT situations by providing scope for examining the influences of cultural demands, past and present intra- and interpersonal factors, and sport and nonsport situational variables. Perhaps the model might help in assessing individual behavior patterns and vulnerability to OT, in anticipating athletes' responses to OT situations, and in understanding athletes' responses to injury and rehabilitation.

CONCLUSIONS

In summary, the experts' and athletes' stories and the OT Risks and Outcomes Model provide detail for enhanced understanding of interactions within the multiple sport and nonsport systems that Meyers and Whelan outlined, and within the physiological, psychological, and social components of OT that Kenttä and Hassmén described. This book, and the new model, present myriad variables that can influence and upset athletes' stress–recovery balances, as well as descriptions of *how* these variables act upon athletes' lives (e.g., by motivating extra training, or by affecting recovery needs). The stories told here, along with the OT Risks and Outcomes Model, also supply information about *why* athletes might be driven to upset their stress–recovery balances, and they provide a background context for athletes' beliefs and behaviors. Furthermore, the model and the stories emphasize dynamic factors in athletes' interactions with people surrounding them and present examples of times when athletes will continue to overtrain against better knowledge and guidance, along with cases where athletes overtrain for varying reasons. The athlete stories provide a rationale for emphasizing injury as one of the most significant outcomes of OT and for looking at OT consequences in assessing injury. Finally, some of the experts' comments about markers might add to the literature on monitoring and identifying OT, and their comments about preventive measures could be integrated into practical interventions aimed at treating or minimizing OT.

12

AFTERWORD
Where to From Here?

In this short final chapter, we want to discuss further implications (and limitations) of our project in this book, along with issues relating to the methods we used to answer our questions. We designed an investigation that helped us achieve the goals of broadening perspectives on, and knowledge about, overtraining (OT) risk factors. Nonetheless, the methods we used had a few limitations that could be addressed with future inquiry. On the positive side, conducting in-depth interviews with experts and athletes about OT experiences gave us insight into the complex interactions among the various influences that drive athletes to overtrain. With a qualitative approach, we had freedom to explore in depth, with experts and athletes, a large range of issues surrounding OT, and we offered athletes opportunities to talk about some of the deeper meanings of OT for them, including why and how they struggled with OT throughout their careers. In particular, during the relatively unstructured interviews we could delve into some of the psychodynamic influences, abusive and coercive drivers, and damaging behaviors revolving around OT processes, responses, and outcomes for the athletes—aspects that have not been addressed in previous literature. After talking with experts from a number of different fields (coaching, sport physiology, sport psychology, and sports medicine), we felt confident that the stories represented a diverse array of perspectives on OT risk factors. In conversations with athletes from different sports, and with different types of OT, we developed the sense that the experience of some OT phenomena might be nearly universal in sport.

In terms of limitations, we conducted an investigation that was exploratory and involved athletes from elite levels of competitive sport, but was not designed to draw conclusive inferences about OT risks for all populations of athletes. OT phenomena most assuredly occur in sub-elite, recreational, and youth sport populations (e.g., overuse injuries in Little League baseball pitchers), and those populations also need to be studied.

At this stage, the OT Risks and Outcomes Model is descriptive and explanatory and has yet to be tested in applied settings to determine its usefulness. Furthermore, we conducted interviews that were one-off interactions with experts and athletes, during which we relied on retrospective accounts of OT experiences. If we had had more time and resources, we might have tapped into additional OT risk factors by following experts and athletes longitudinally throughout one or more competitive seasons. Thus future research might be directed toward refining the risk factor list and the model itself (see more on possible future directions of research later in this chapter).

CURRENT APPLICATIONS

Over the last several years, Sean has been working with coaches, athletes, dancers, teachers, medical staff, sport and performing arts governing bodies, and corporate and organizational clients, educating them on stress–recovery balance using the OT Risks and Outcomes Model. In the following paragraphs Sean highlights his experiences using the model in applied settings.

One of my first presentations based on the project, and using the model, was to a group of hockey coaches attending a conference in Melbourne, Australia. As I walked into the room, several of the coaches approached me, and one of them said, "Look. We don't have any problems with overtraining in our sport, so I am not sure what we are going to get out of this." I responded with, "I think I have some sense of where you are coming from because I said similar things to myself several times during the course of my research. It would be great if you could stick around for the seminar because your experiences would be a valuable contribution to what we are going to talk about. Besides, I would love to see you walk out of here with another perspective on this whole 'overtraining' thing!" The group of coaches decided to stick around, and they ended up being my biggest supporters in the seminar. What I realized in bringing the model to the applied world is that many athletes, coaches, and others see overtraining as one-dimensional—an outcome of too much training that leads to a chronic fatigue type of syndrome. My role in applied settings has become one of educating people about the multidimensional aspects of stress and recovery, the multifaceted risk factors for stress–recovery imbalances, and the multiple adverse outcomes of overtraining behaviors and underrecovery. The key point in application of the model has been to broaden the focus on sport factors (training and competition) as the only sources of stress from which one needs recovery, to include nonsport stressors, and to highlight holistic approaches to recovery that are needed in response to these stressors.

In helping athletes and other performers understand the risk factors for stress–recovery imbalances, I have found that they become better equipped to make effective decisions about managing their fatigue levels and coping with minor injuries, illnesses, and time off from training. With increased understanding, athletes and other performers also seem better equipped to make more effective decisions during rehabilitation and upon their return to activities after injury. Furthermore, I have found effective applications of the model in less physically demanding sports, such as golf and shooting, and in other areas of performance, such as ballet, music, and business, where people can experience high levels of fatigue and increased rates of injury and illness, whether from lengthy periods of concentration, overuse of specific body parts, or stress (and distress) in what is often referred to as work–life balance.

Although many organizations have approached me looking for quick fixes to ongoing injury, fatigue, and illness issues, the education about stress–recovery balance using our model has become, for some, the starting point for cultural change. Athletes and others are still going to become sick, tired, and injured—even in transformed cultures—but for those dedicated to holistic understanding of, and responses to, stress–recovery imbalances, the injury and illness rates will drop over time, and the decision making about response to first signs of imbalance will improve, such that prevention, rather than treatment, can become the emphasis.

FUTURE DIRECTIONS

We hope this project stimulates research in a number of directions and real-world applications. With the emphasis on injury, illness, and psychological distress as significant outcomes of OT processes, we think that research could be directed at more holistic evaluations of OT-related injuries, illnesses, and mental health concerns. In particular, researchers could conduct longitudinal studies directed at examining injuries, illness, and psychosocial consequences within the context of OT. With such studies, researchers could illuminate more OT-related causes of injury, illness, and mental health problems and perhaps provide insight into the types and frequencies of these issues associated with OT. Researchers involved in such studies could take both case study and survey approaches to gathering information about injury, illness, and psychosocial problems in OT contexts. Case studies could provide additional perspectives on the idiosyncratic aspects of injury, illness, and psychosocial issues for different athletes, whereas survey research could provide data about types and frequencies of injury, illness, and psychosocial problems associated with OT behaviors. With any research, we hope that investigators would include measures of performance—the most important outcomes for coaches and athletes—to compare with the other findings of their studies.

Much of the past research in OT has taken a quantitative group analysis approach. We feel that this type of research has limitations in that, when individual lives and experiences are homogenized into group means and group-mean changes over time, individual differences disappear and are lost in the statistical analysis. OT is a highly idiosyncratic process. For example, Martin, Andersen, and Gates (2000) found that high-intensity training led to more mood disturbance (OT-related symptom) in cyclists, but that when they examined the data at an individual level, the relationships between mood disturbance and the ultimate variable of interest, performance after taper, were all over the place. Some highly mood-disturbed cyclists performed excellently after taper, and some mildly mood-disturbed cyclists had poor performances. Group data say one thing; individual data may tell different stories. We are not saying quantitative methods should not be used. They have immense value, but that value lies more in the epidemiological realm than in the intimate landscape of individual lives (see suggestions for epidemiological prevalence studies later in this chapter).

In addition to exploring the multiple outcomes of OT, future researchers could continue the exploration into OT risk factors that we initiated in this book. As with injury and illness studies, researchers could conduct both case study and survey research on OT risk factors to augment the current investigations. Researchers could carry out case studies in which they monitor OT risk in athletes over the course of a season, or longer, and could develop and administer risk assessment surveys, derived from the risk factor list and the OT Risks and Outcomes Model. Researchers conducting case study and survey research could explore such issues as athletes', parents', and coaches' attitudes toward, and behaviors associated with, performance, success, failure, training, recovery, injury, illness, psychological distress, and other OT processes and outcomes outlined in the model. Furthermore, researchers could develop an applied checklist of OT risk factors, which could provide quantitative measures of risk to be compared with other statistics gathered from performance measures, quantifiable physiological and psychological markers of OT (if any reliable markers are established), and injury and illness statistics. In conducting idiographic research and large-scale survey studies, researchers could provide greater understanding of both the specific experiences of, and general trends in, OT risk, and of the attitudes and beliefs regarding OT that people hold across multiple sports and cultures.

Looking at detailed descriptions of OT from the current model, future researchers could also conduct more precise epidemiological prevalence investigations than have been done in the past (e.g., Morgan, O'Connor, Ellickson, & Bradley, 1988; Morgan, O'Connor, Sparling, & Pate, 1987), research that was based on vague definitions of OT. In conducting OT prevalence studies, researchers could gather more specific and meaningful data about the occurrence of OT signs, processes, and outcomes than have been documented in past research. Researchers could also develop and administer questionnaires, based on the model, which present a series of specific questions for athletes regarding the frequency of OT processes and outcomes (e.g., questions about time and intensity of exercise per week outside of coach-sanctioned training, or number of days per week of restricted caloric intake).

Prevention and Treatment of OT

Although prevention and treatment were not focal points of this book, researchers could use the stories and model presented here to provide a framework for exploring potential OT interventions. The experts talked about OT education and awareness as important factors in prevention and pointed out the significance of improving communication patterns between coaches, athletes, and parents. The athletes' stories illustrated how a lack of education and awareness about OT issues, as well as poor communication and maladaptive reinforcement patterns, greatly influenced athletes' OT behaviors. Researchers could conduct studies that test the effects of communication skills and educational interventions centered on OT risk factors for coaches, parents, and athletes. Researchers could also conduct studies in which they apply the risk factor list and the OT Risks and Outcomes Model to identifying OT risks in athletes, then use preventive techniques and interventions aimed at minimizing risks and optimizing stress recovery balance. Researchers could apply these intervention studies on a longitudinal basis, during which comparison groups could be given staggered intervention protocols.

Finally, in keeping with the problems expressed by Trisha Leahy, sociological and ethnographic studies of pathogenic sport cultures and institutions would help illuminate how systems of abuse are perpetuated, how exploitation of athlete vulnerabilities is endemic to sport, and how we may see people (mistakenly) as means to certain ends (e.g., more medals, fame for coaches), rather than ends in themselves.

Implications for Professional Practice

Applied practitioners in a number of sport and health professions—such as physicians, physiotherapists, psychologists, exercise physiologists, athletic trainers, and coaches—could apply the risk factor list and the OT Risks and Outcomes Model to the following areas:

- Education of athletes, coaches, parents, health professionals, and sport administrators about OT risks and outcomes, especially about the not-so-obvious risk factors and outcomes depicted in the model
- Development and implementation of OT assessment and monitoring for the purpose of minimizing OT and its outcomes
- Exploration and identification of athletes' OT experiences within psychological or medical health consultations to enhance therapies and interventions

First, practitioners could incorporate the results of this investigation into education seminars for athletes, coaches, parents, health professionals, and sport administrators. To help drive behavior changes that reduce the prevalence of OT, people need to become aware of the myriad variables—intrapersonal, interpersonal, situational, and sociocultural—that might contribute to OT processes and outcomes. Athletes could benefit from increased education in that they might improve self-monitoring behaviors for OT risk, whereas coaches and parents could benefit from learning more constructive reinforcement patterns, thus being better positioned to offer balanced guidance. Health professionals might be better equipped to provide accurate medical and psychological diagnoses, conduct thorough clinical assessments, and plan effective rehabilitation and recovery activities. Sport administrators might be more likely to initiate beneficial systemic, political, and cultural changes with regard to OT and its outcomes.

Second, experts, coaches, athletes, parents, and others could use the material in this book to help them identify OT risk during its early stages, when steps can still be taken to minimize it. People working with athletes could apply the following suggestions about prevention from the experts in chapter 4:

1. Engaging in preventive actions and behaviors
   - Take initiative to reduce training, if necessary.
   - Monitor athletes' feelings regarding performance.
   - Monitor and assess athletes' beliefs about training and recovery.
   - Monitor physical and psychological signals regarding fatigue and recovery.
   - Design training programs according to individual athlete needs.
   - Monitor stressors in athletes' lives, both inside and outside of sport.
   - Try not to sustain, unrealistically, an athlete's peak in performance.

2. Improving education and awareness
   - Help athletes develop awareness of different levels of fatigue.
   - Develop awareness of one's limitations as a coach.
   - Educate parents, coaches, and athletes regarding abuse and personal boundary issues.
   - Educate athletes, coaches, parents, and others regarding recovery and other issues related to a balanced approach to life.
   - Educate athletes about individual differences in training and recovery needs, emphasizing that "one size does not fit all" (or even many) athletes.
   - Emphasize athletes' life balance.

3. Enhancing communication
   - Communicate about the importance of rest and recovery.
   - Emphasize open communication to and from athletes, especially regarding injury, fatigue, illness, psychological distress, and other life stressors.

Finally, psychologists, or others working with athletes in therapeutic settings, could use the model and the stories as a framework to explore and identify the meanings of OT experiences for athletes as they help athletes achieve general life balance in both sport and nonsport contexts. Practitioners, such as physicians and psychologists, might refer to the risk factor list and the OT Risks and Outcomes Model to explore causes of maladaptive behaviors and poor performances, which could help them to guide change involving such behaviors. Practitioners could also use this book to help in the assessment of, responses to, and rehabilitation from injury, and in making more precise attributions about the causes of injury and more effective plans for future injury prevention.

PIPE DREAMS

We three are a hopeful bunch, but we are also realists. Our optimism about early recognition of OT signs and symptoms and the prevention of future damage to athletes is tempered by the realities of sport traditions, abusive environments, and resistance to change. We may use the OT Risks and Outcomes Model as an educational tool at the individual level with some athletes and coaches, and may even effect some healthy changes. But how do we change the culture those coaches and athletes operate in? What good is it to educate and train if that education and training come up against monolithic traditions that work for the few (the champions), but leave trails of sadness, disappointment, alienation, and broken bodies and spirits? Our David seems tiny, and the Goliath of sport culture appears invulnerable.

For example, Sean (see chapter 10) was armed with knowledge, training, and a keen understanding of OT signs and symptoms, and he knew what he needed to do to stay healthy. He returned to the world of Olympic rowing, and in that environment, coupled with the irrational desires of his heart, he fell into the demands of the culture and damaged himself. If an expert like Sean succumbs, then what does that say about the power of the systems in sport that hold such sway over athletes and coaches?

We still believe in education, and getting our message out there to athletes, coaches, national governing bodies, and—maybe most important—our students. They will be the ones to carry on the fight to help make sport a place for demonstrating competence (at all levels), for experiencing joy of functioning, and for healing rifts in mind, body, and spirit. Sport is not always such a place, but we three authors have a softness for what look like noble but challenging battles. Maybe in the future, with more education, and more stories of damaged lives, we will see David grow and Goliath shrink a bit. As we stated, we are hopeful.

PARTING GLANCES

Much of this book is about personal journeys. Sean's project was itself a personal journey for all of us. Developing, executing, and completing such a project involves a relatively long-term intimate relationship between the student and his research advisors. That relationship is fraught with many of the aspects of other long-term interpersonal interactions. There is joy, frustration, excitement, anger, hope, pleasure, pain, and fascination. And that's just a start to the list of what we have experienced. It was not all sweetness and light. There were times when Sean wanted to hire a hit man to take care of Mark. To quote the Grateful Dead, "What a long, strange trip it's been." With all the stumbles, delays, and anxieties, and with all the "Wow!" moments and "Damn! This is good!" times, we think we came through the journey with something of value, both for us as professionals trying to make a difference, and for the sport community.

And the journey is not over. It never really is.

Suggested Further Readings

Achten, J., & Jeukendrup, A.E. (2003). Heart rate monitoring: Applications and limitations. *Sports Medicine, 33*, 517–538.

Benhaddad, A.A., Bouix, D., Khaled, S., Micallef, J.P., Mercier, J., Bringer, J., & Brun, J.F. (1999). Early hemorheologic aspects of overtraining in elite athletes. *Clinical Hemorheology & Microcirculation, 20*, 117–125.

Brewer, B.W., & Petrie, T.A. (2002). Psychopathology in sport and exercise. In J.L. Van Raalte & B.W. Brewer (Eds.), *Exploring sport and exercise psychology* (2nd ed., pp. 307–323). Washington, DC: American Psychological Association.

Dale, J., & Weinberg, R. (1990). Burnout in sport: A review and critique. *Journal of Applied Sport Psychology, 2*, 67–83.

Davis, H., IV, Botterill, C., & MacNeill, K. (2002). Mood and self-regulation changes in underrecovery: An intervention model. In M. Kellmann (Ed.), *Enhancing recovery: Preventing underperformance in athletes* (pp. 161–180). Champaign, IL: Human Kinetics.

Evans, B. K., & Fischer, D. (1993). The nature of burnout: A study of the three-factor model of burnout in human services and non-human service samples. *Journal of Occupational & Organizational Psychology, 66*, 29–38.

Glaser, B.G. (1992). *Basics of grounded theory analysis*. Mill Valley, CA: Sociology Press.

Glaser, B.G., & Strauss, A.L. (1979). *The discovery of grounded theory: Strategies for qualitative research*. New York: Aldine.

Gould, D., & Dieffenbach, K. (2002). Overtraining, underrecovery, and burnout in sport. In M. Kellmann (Ed.), *Enhancing recovery: Preventing underperformance in athletes* (pp. 25–36). Champaign, IL: Human Kinetics.

Gould, D., Eklund, R.C., & Jackson, S.A. (1993). Coping strategies used by U.S. Olympic wrestlers. *Research Quarterly for Exercise and Sport, 64*, 78–93.

Gould, D., Jackson, S.A., & Finch, L.M. (1993). Sources of stress in national champion figure skaters. *Journal of Sport & Exercise Psychology, 15*, 134–159.

Gould, D., Tuffey, S., Udry, E., & Loehr, J. (1996). Burnout in competitive junior tennis players: II. Qualitative analysis. *The Sport Psychologist, 10*, 341–366.

Gould, D., Udry, E., Tuffey, S., & Loehr, J. (1996). Burnout in competitive junior tennis players: I. A quantitative psychological assessment. *The Sport Psychologist, 10*, 322–340.

Greenleaf, C., Gould, D., & Dieffenbach, K. (2001). Factors influencing Olympic performance: Interviews with Atlanta and Nagano U.S. Olympians. *Journal of Applied Sport Psychology, 13*, 154–184.

Guilliat, R. (2005, September 17). Once were warriors: The painful price of sporting glory. *Good Weekend: The Age Magazine*, 26–32.

Hardy, L., Jones, G., & Gould, D. (1996). *Understanding psychological preparation for sport: Theory and practice for elite performers*. West Sussex, England: Wiley.

Hogg, J.M. (2002). Debriefing: A means to increasing recovery and subsequent performance. In M. Kellmann (Ed.), *Enhancing recovery: Preventing underperformance in athletes* (pp. 181–198). Champaign, IL: Human Kinetics.

Jackson, S.E., Schwab, R.L., & Schuler, R.S. (1986). Toward an understanding of the burnout phenomenon. *Journal of Applied Psychology, 71*, 630–640.

Kelley, B.C. (1994). A model of stress and burnout in collegiate coaches: Effects of gender and time of season. *Research Quarterly for Exercise and Sport, 65*, 48–58.

Kelley, B., & Eklund, R.C. (1999). A model of stress and burnout in male high school athletic directors. *Journal of Sport & Exercise Psychology, 21*, 280–294.

Kelley, B.C., & Gill, D.L. (1993). An examination of personal/situational variables, stress appraisal, and burnout in collegiate teacher–coaches. *Research Quarterly for Exercise and Sport, 64*, 94–102.

Kellmann, M., & Kallus, K. W. (1999). Mood, recovery–stress state, and regeneration. In M. Lehmann, C. Foster, U. Gastmann, H. Keizer, & J. M. Steinacker (Eds.), *Overload, performance incompetence, and regeneration in sport* (pp. 101–117). New York: Plenum.

Kellmann, M., Patrick, T., Bottrill, C., & Wilson, C. (2002). The recovery–cue and its use in applied settings: Practical suggestions regarding assessment and monitoring of recovery. In M. Kellmann (Ed.), *Enhancing recovery: Preventing underperformance in athletes* (pp. 219–232). Champaign, IL: Human Kinetics.

Lee, R.T., & Ashforth, B.E. (1990). On the meaning of Maslach's three dimensions of burnout. *Journal of Applied Psychology, 75*, 743–747.

Lehmann, M., Foster, C., Gastmann, U., Keizer, H.A., & Steinacker, J.M. (1999). Definitions, types, symptoms, findings, underlying mechanisms, and frequency of overtraining and overtraining syndrome. In M. Lehmann, C. Foster, U. Gastmann, H. Keizer, & J.M. Steinacker

(Eds.), *Overload, performance incompetence, and regeneration in sport* (pp. 1–6). New York: Plenum.

Martens, R. (1987). Science, knowledge, and sport psychology. *The Sport Psychologist, 1,* 29–55.

Martin, J.J., Kelley, B., & Eklund, R.C. (1999). A model of stress and burnout in male high school athletic directors. *Journal of Sport & Exercise Psychology, 21,* 280–294.

Meier, S.T. (1984). The construct validity of burnout. *Journal of Occupational Psychology, 57,* 211–219.

Miles, M.B., & Huberman, A.M. (1994). *Qualitative data analysis.* Thousand Oaks, CA: Sage.

Norris, S.R., & Smith, D.J. (2002). Planning, periodization, and sequencing of training and competition: The rationale for a competently planned, optimally executed training and competition program, supported by a multidisciplinary team. In M. Kellmann (Ed.) *Enhancing recovery: Preventing underperformance in athletes* (pp. 121–142). Champaign, IL: Human Kinetics.

Patton, M.Q. (1990) *Qualitative evaluation and research methods.* Newbury Park, CA: Sage.

Posig, M., & Kickul, J. (2003). Extending our understanding of burnout: Test of an integrated model in nonservice occupations. *Journal of Occupational Health Psychology, 8,* 3–19.

Raedeke, T.D. (1997). Is athlete burnout more than just stress? A sport commitment perspective. *Journal of Sport & Exercise Psychology, 19,* 396–417.

Raedeke, T.D., & Smith, A.L. (2001). Development and preliminary validation of an athlete burnout measure. *Journal of Sport & Exercise Psychology, 23,* 281–306.

Rainey, D.W. (1995). Stress, burnout, and intention to terminate among umpires. *Journal of Sport Behavior, 18,* 312–323.

Schmidt, G.W., & Stein, G.L. (1991). Sport commitment: A model integrating enjoyment, dropout, and burnout. *Journal of Sport & Exercise Psychology, 13,* 254–265.

Smith, R.E. (1986). Toward a cognitive–affective model of athletic burnout. *Journal of Sport Psychology, 8,* 36–50.

Smith, R.E. (1988). The logic and design of case study research. *The Sport Psychologist, 2,* 1–12.

Strauss, A.L., & Corbin, J. (1990). *Basics of qualitative research.* Newbury Park, CA: Sage.

Udry, E., Gould, D., Bridges, D., & Tuffey, S. (1997). People helping people? Examining the social ties of athletes coping with burnout and injury stress. *Journal of Sport & Exercise Psychology, 19,* 368–395.

Vealey, R.S., Armstrong, L., & Comar, W. (1998). Influence of perceived coaching behaviors on burnout and competitive anxiety in female college athletes. *Journal of Applied Sport Psychology, 10,* 297–318.

Vealey, R.S., Udry, E.M., Zimmerman, V., & Soliday, J. (1992). Intrapersonal and situational predictors of coaching burnout. *Journal of Sport & Exercise Psychology, 14,* 40–58.

Wolcott, H. F. (1990). *Writing up qualitative research.* Newbury Park, CA: Sage.

References

Introduction

Kenttä, G., & Hassmén, P. (2002). Underrecovery and overtraining: A conceptual model. In M. Kellmann (Ed.), *Enhancing recovery: Preventing underperformance in athletes* (pp. 57–80). Champaign, IL: Human Kinetics.

Sparkes, A.C. (2002). *Telling tales in sport and physical activity: A qualitative journey.* Champaign, IL: Human Kinetics.

Chapter 1

Armstrong, L.E., & VanHeest, J.L. (2002). The unknown mechanism of the overtraining syndrome: Clues from depression and psychoneuroimmunology. *Sports Medicine, 32,* 185–209.

Bompa, T. (1983). *Theory and methodology of training: The key to athletic performance.* Dubuque, IA: Kendall/Hunt.

Budgett, R. (1998). Fatigue and underperformance in athletes: The overtraining syndrome. *British Journal of Sports Medicine, 32,* 107–110.

Freudenberger, H. (1974). Staff burn-out. *Journal of Social Issues, 30,* 159–165.

Gould, D., Guinan, D., Greenleaf, C., Medbery, R., & Peterson, K. (1999). Factors affecting Olympic performance: Perceptions of athletes and coaches from more and less successful teams. *The Sport Psychologist, 13,* 371–394.

Gould, D., Tuffey, S., Udry, E., & Loehr, J. (1997). Burnout in competitive junior tennis players: III. Individual differences in the burnout experience. *The Sport Psychologist, 11,* 257–276.

Griffith, C.R. (1926). *Psychology of coaching.* New York: Scribner's.

Hackney, A.C., Pearman, S.N., & Nowacki, J.M. (1990). Physiological profiles of overtrained and stale athletes: A review. *Journal of Applied Sport Psychology, 2,* 21–33.

Hawley, C.J., & Schoene, R. (2003). Overtraining syndrome. *Physician and Sportsmedicine, 31*(6), 25–31.

Hooper, S.L., & Mackinnon, L.T. (1995). Monitoring overtraining in athletes. *Sports Medicine, 20,* 321–327.

Kellmann, M. (2002). Underrecovery and overtraining: Different concepts—similar impact? In M. Kellmann (Ed.), *Enhancing recovery: Preventing underperformance in athletes* (pp. 3–24). Champaign, IL: Human Kinetics.

Kenttä, G., & Hassmén, P. (2002). Underrecovery and overtraining: A conceptual model. In M. Kellmann (Ed.), *Enhancing recovery: Preventing underperformance in athletes* (pp. 57–80). Champaign, IL: Human Kinetics.

Krane, V., Greenleaf, C.A., & Snow, J. (1997). Reaching for gold and the price of glory: A motivational case study of an elite gymnast. *The Sport Psychologist, 11,* 53–71.

Kreider, R.B., Fry, A.C., & O'Toole, M.L. (1998). Overtraining in sport: Terms, definitions, and prevalence. In R.B. Kreider, A.C. Fry, & M.L. O'Toole (Eds.), *Overtraining in sport* (pp. vii–ix). Champaign, IL: Human Kinetics.

Lehmann, M., Foster, C., Gastmann, U., Keizer, H. A., & Steinacker, J.M. (1999). Definitions, types, symptoms, findings, underlying mechanisms, and frequency of overtraining and overtraining syndrome. In M. Lehmann, C. Foster, U. Gastmann, H. Keizer, & J.M. Steinacker (Eds.), *Overload, performance incompetence, and regeneration in sport* (pp. 1–6). New York: Plenum.

Lehmann, M., Foster, C., & Keul, J. (1993). Overtraining in endurance athletes: A brief review. *Medicine & Science in Sports & Exercise, 25,* 854–862.

Martin, D.T., Andersen, M.B., & Gates, W. (2000). Using profile of mood states (POMS) to monitor high-intensity training in cyclists: Groups versus case studies. *The Sport Psychologist, 14,* 138–156.

Maslach, C. (1982). *Burnout: The cost of caring.* New York: Prentice Hall.

Maslach, C., & Jackson, S.E. (1981). *The Maslach Burnout Inventory.* Palo Alto, CA: Consulting Psychologists Press.

McCann, S. (1995). Overtraining and burnout. In S.M. Murphy (Ed.), *Sport psychology interventions* (pp. 347–369). Champaign, IL: Human Kinetics.

Meeusen, R. (2006). Prevention, diagnosis and treatment of the overtraining syndrome. *European Journal of Sport Science, 6,* 1–14.

Meyers, A.W., & Whelan, J.P. (1998). A systemic model for understanding psychological influences in overtraining. In R.B. Kreider, A.C. Fry, & M.L. O'Toole (Eds.), *Overtraining in sport* (pp. 335–372). Champaign, IL: Human Kinetics.

Nederhof, E., Lemmink, K., Visscher, C., Meeusen, R., & Mulder, T. (2006). Psychomotor speed: Possibly a new marker for overtraining. *Sports Medicine, 36,* 817–828.

O'Toole, M.L. (1998). Overreaching and overtraining in endurance athletes. In R.B. Kreider, A.C. Fry, & M.L. O'Toole (Eds.), *Overtraining in sport* (pp. 3–18). Champaign, IL: Human Kinetics.

Parmenter, D.C. (1923). Some medical aspects of the training of college athletes. *The Boston Medical and Surgical Journal, 189,* 45–50.

Peterson, K. (2005). Overtraining: Balancing practice and performance. In S. Murphy (Ed.), *The sport psych handbook* (pp. 49–70). Champaign, IL: Human Kinetics.

Raglin, J.S. (1993). Overtraining and staleness: Psychometric monitoring of endurance athletes. In R.N. Singer, M. Murphey, & L.K. Tennant (Eds.), *Handbook of research on sport psychology* (pp. 840–850). New York: Macmillan.

Raglin, J.S., & Wilson, G.S. (2000). Overtraining in athletes. In Y. L. Hanin (Ed.), *Emotions in sport* (pp. 191–207). Champaign, IL: Human Kinetics.

Rowbottom, D.G., Keast, D., & Morton, A.R. (1998). Monitoring and preventing of overreaching and overtraining in endurance athletes. In R.B. Kreider, A.C. Fry, & M.L. O'Toole (Eds.), *Overtraining in sport* (pp. 47–66). Champaign, IL: Human Kinetics.

Silva, J.M., III. (1990). An analysis of the training stress syndrome in competitive athletics. *Journal of Applied Sport Psychology, 2,* 5–20.

Steinacker, J.M., & Lehmann, M. (2002). Clinical findings and mechanisms of stress and recovery in athletes. In M. Kellmann (Ed.), *Enhancing recovery: Preventing underperformance in athletes* (pp. 103–118). Champaign, IL: Human Kinetics.

Urhausen, A., & Kindermann, W. (2002). Diagnosis of overtraining: What tools do we have? *Sports Medicine, 32,* 95–102.

Uusitalo, A.L. (2001). Overtraining: Making a difficult diagnosis and implementing targeted treatment. *Physician and Sportsmedicine, 29*(5), 35–50.

Uusitalo, A.L. (2006). A comment on: Prevention, diagnosis and treatment of the overtraining syndrome. *European Journal of Sport Science, 6,* 261–262.

Chapter 2

Armstrong, L.E., & VanHeest, J.L. (2002). The unknown mechanism of the overtraining syndrome: Clues from depression and psychoneuroimmunology. *Sports Medicine, 32,* 185–209.

Berglund, B., & Säfström, H. (1994). Psychological monitoring and modulation of training load of world–class canoeists. *Medicine & Science in Sports & Exercise, 26,* 1036–1040.

Callister, R., Callister, R.J., Fleck, S.J., & Dudley, G.A. (1990). Physiological and performance responses to overtraining in elite judo athletes. *Medicine & Science in Sports & Exercise, 22,* 816–824.

Chicharro, J.L., Lopezmojares, L.M., Lucia, A., Perez, M., Alvarez, J., Labanda, P., Calvo, F., & Vaquero, A.F. (1998). Overtraining parameters in special military units. *Aviation Space & Environmental Medicine, 69,* 562–568.

Costill, D.L. (1986). *Inside running: Basics of sport physiology.* Indianapolis, IN: Benchmark.

Flynn, M. G. (1998). Future research needs and directions. In R. B. Kreider, A. C. Fry, & M. L. O'Toole (Eds.), *Overtraining in sport* (pp. 373–383). Champaign, IL: Human Kinetics.

Fry, A.C. (1998). The role of training intensity in resistance exercise overtraining and overreaching. In R.B. Kreider, A.C. Fry, & M.L. O'Toole (Eds.), *Overtraining in sport* (pp. 107–127). Champaign, IL: Human Kinetics.

Fry, A.C., Kraemer, W.J., Lynch, J.M., Triplett, N.T, & Koziris, L.P. (1994). Does short–term near–maximal intensity machine resistance exercise induce overtraining? *Journal of Applied Sport Science Research, 8,* 188–191.

Fry, A.C., Kraemer, W.J., van Borselen, F., Lynch, J.M., Triplett, N.T., Koziris, L.P., & Fleck, S.J. (1994). Catecholamine responses to short–term high–intensity resistance exercise overtraining. *Journal of Applied Physiology, 77,* 941–946.

Fry, R.W., Morton, A.R., & Keast, D. (1991). Overtraining in athletes: An update. *Sports Medicine, 12,* 32–65.

Gleeson, M., McDonald, W.A., Pyne, D.B., Clancy, R.I., Cripps, A.W., Francis, J.L., & Fricker, P.A. (2000). Immune status and respiratory illness for elite swimmers during a 12–week training cycle. *International Journal of Sports Medicine, 21,* 302–307.

Gotovtseva, E.P., Surkina, I.D., & Uchakin, P.N. (1998). Potential interventions to prevent immunosuppression during training. In R.B. Kreider, A.C. Fry, & M.L. O'Toole (Eds.), *Overtraining in sport* (pp. 243–272). Champaign, IL: Human Kinetics.

Gould, D., Greenleaf, C., Chung, Y., & Guinan, D. (2002). A survey of U.S. Atlanta and Nagano Olympians: Variables perceived to influence performance. *Research Quarterly for Exercise and Sport, 73,* 175–186.

Halson, S.L., Lancaster, G.I., Jeukendrup, A.E., & Gleeson, M. (2003). Immunological responses to overreaching in cyclists. *Medicine & Science in Sports & Exercise, 35,* 854–861.

Hedelin, R., Kenttä, G., Wiklund, U., Bjerle, P., & Henriksson–Larsén, K. (2000). Short–term overtraining: Effects on performance, circulatory responses, and heart rate variability. *Medicine & Science in Sports & Exercise, 32,* 1480–1484.

Hooper, S.L., & Mackinnon, L.T. (1995). Monitoring overtraining in athletes. *Sports Medicine, 20,* 321–327.

Hooper, S.L., Mackinnon, L.T., & Hanrahan, S. (1997). Mood states as an indication of staleness and recovery. *International Journal of Sport Psychology, 28,* 1–12.

Hooper, S.L., Mackinnon, L.T., & Howard, A. (1999). Physiological and psychometric variables for monitoring recovery during tapering for major competition. *Medicine & Science in Sports & Exercise, 31*, 1205–1210.

Hooper, S.L., Mackinnon, L.T., Howard, A., Gordon, R.D., & Bachmann, A.W. (1995). Markers for monitoring overtraining and recovery. *Medicine & Science in Sports & Exercise, 27*, 106–112.

Jeukendrup, A.E., & Hesselink, M.K.C. (1994). What do lactate curves tell us? *British Journal of Sports Medicine, 28*, 239–240.

Jokl, E. (1974). The immunological status of athletes. *Journal of Sports Medicine, 14*, 165–167.

Kaplan, R.M. (1990). Behavior as the central outcome in health care. *American Psychologist, 45*, 1211–1220.

Kellmann, M. (2002). Underrecovery and overtraining: Different concepts—similar impact? In M. Kellmann (Ed.), *Enhancing recovery: Preventing underperformance in athletes* (pp. 3–24). Champaign, IL: Human Kinetics.

Kellmann, M., Altenburg, D., Lormes, W., & Steinacker, J.M. (2001). Assessing stress and recovery during preparation for the World Championships in rowing. *The Sport Psychologist, 15*, 151–167.

Kellmann, M., & Günther, K.D. (2000). Changes in stress and recovery in elite rowers during preparation for the Olympic Games. *Medicine & Science in Sports & Exercise, 32*, 676–683.

Kellmann, M., & Kallus, K.W. (2001). *Recovery–Stress Questionnaire for Athletes: User manual.* Champaign, IL: Human Kinetics.

Kenttä, G., & Hassmén, P. (1998). Overtraining and recovery: A conceptual model. *Sports Medicine, 26*, 1–16.

Kenttä, G., & Hassmén, P. (2002). Underrecovery and overtraining: A conceptual model. In M. Kellmann (Ed.), *Enhancing recovery: Preventing underperformance in athletes* (pp. 57–80). Champaign, IL: Human Kinetics.

Kenttä, G., Hassmén, P., & Raglin, J.S. (2001). Training practices and overtraining syndrome in Swedish age–group athletes. *International Journal of Sports Medicine, 22*, 460–465.

Kibler, B.W., & Chandler, T.J. (1998). Musculoskeletal and orthopedic considerations. In R.B. Kreider, A.C. Fry, & M.L. O'Toole (Eds.), *Overtraining in sport* (pp. 169–190). Champaign, IL: Human Kinetics.

Kuipers, H. (1996). How much is too much? Performance aspects of overtraining. *Research Quarterly for Exercise and Sport, 67*(Suppl. 3), S65–S69.

Lehmann, M., Baur, S., Netzer, N., & Gastmann, U. (1997). Monitoring high–intensity endurance training using neuromuscular excitability to recognize overtraining. *European Journal of Applied Physiology & Occupational Physiology, 76*, 187–191.

Lehmann, M., Dickhuth, H.H., & Gendrisch, G. (1991). Training—overtraining: A prospective, experimental study with experienced middle– and long–distance runners. *International Journal of Sports Medicine, 12*, 444–452.

Lehmann, M., Foster, C., & Keul, J. (1993). Overtraining in endurance athletes: A brief review. *Medicine & Science in Sports & Exercise, 25*, 854–862.

Lehmann, M., Foster, C., Netzer, N., Lormes, W., Steinacker, J.M., Lui, Y., Optiz–Gress, A., & Gastmann, U. (1998). Physiological responses to short– and long–term overtraining in endurance athletes. In R.B. Kreider, A.C. Fry, & M.L. O'Toole (Eds.), *Overtraining in sport* (pp. 19–46). Champaign, IL: Human Kinetics.

Mackinnon, L.T. (1998). Effects of overreaching and overtraining on immune function. In R.B. Kreider, A.C. Fry, & M.L. O'Toole (Eds.), *Overtraining in sport* (pp. 219–241). Champaign, IL: Human Kinetics.

Mackinnon, L.T., & Hooper, S. (1994). Mucosal (secretory) immune system responses to exercise of varying intensity and during overtraining. *International Journal of Sports Medicine, 15*(Suppl. 3), S179–S183.

Mackinnon, L.T., Hooper, S.L., Jones, S., Gordon, R.D., & Bachmann, A.W. (1997). Hormonal, immunological, and haematological responses to intensified training in elite swimmers. *Medicine & Science in Sports & Exercise, 29*, 1637–1645.

Martin, D.T., Andersen, M.B., & Gates, W. (2000). Using Profile of Mood States (POMS) to monitor high–intensity training in cyclists: Groups versus case studies. *The Sport Psychologist, 14*, 138–156.

McNair, D.M., Lorr, M., & Droppleman, L.F. (1992). *POMS manual. Profile of Mood States.* San Diego: Educational and Industrial Testing Service.

Meyers, A.W., & Whelan, J.P. (1998). A systemic model for understanding psychological influences in overtraining. In R.B. Kreider, A.C. Fry, & M.L. O'Toole (Eds.), *Overtraining in sport* (pp. 335–372). Champaign, IL: Human Kinetics.

Morgan, W.P., Brown, D.R., Raglin, J.S., O'Connor, P.J., & Ellickson, K.A. (1987). Physiological monitoring of overtraining and staleness. *British Journal of Sports Medicine, 21*, 107–114.

Morgan, W.P., Costill, D.L., Flynn, M.G., Raglin, J.S., & O'Connor, P.J. (1988). Mood disturbances following increased training in swimmers. *Medicine & Science in Sports & Exercise, 20*, 408–414.

Morgan, W.P., O'Connor, P.J., Ellickson, K.A., & Bradley, P.W. (1988). Personality structure, mood states, and performance in elite male distance runners. *International Journal of Sport Psychology, 19*, 247–263.

Morgan, W.P., O'Connor, P.J., Sparling, P.B., & Pate, R.R. (1987). Psychological characterization of the elite female distance runner. *International Journal of Sports Medicine, 8*, 124–131.

Morton, R.H. (1997). Modelling training and overtraining. *Journal of Sports Sciences, 15*, 335–340.

Murphy, S.M., Fleck, S.J., Dudley, G., & Callister, R. (1990). Psychological and performance concomitants of increased volume training in elite athletes. *Journal of Applied Sport Psychology, 2,* 34–50.

Naessens, G., Chandler, T.J., Kibler, W.B., & Driessens, M. (2000). Clinical usefulness of nocturnal urinary noradrenaline excretion patterns in the follow–up of training processes in high–level soccer players. *Journal of Strength and Conditioning Research, 14,* 125–131.

Nieman, D.C. (1997). Risk of upper respiratory tract infection in athletes: An epidemiologic and immunologic perspective. *Journal of Athletic Training, 32,* 344–349.

Nieman, D.C. (1998). Effects of athletic endurance training on infection rates and immunity. In R.B. Kreider, A.C. Fry, & M.L. O'Toole (Eds.), *Overtraining in sport* (pp. 193–217). Champaign, IL: Human Kinetics.

Nieman, D.C. (2000). Prevention of upper respiratory tract infection in endurance athletes. *International SportMed Journal, 1*(2), 1–6.

Nieman, D.C., & Pedersen, B.K. (1999). Exercise and immune function: Recent developments. *Sports Medicine, 27,* 73–80.

O'Connor, P.J. (1998). Overtraining and staleness. In W.P. Morgan (Ed.), *Physical activity and mental health* (pp. 145–160). New York: Hemisphere.

O'Toole, M.L. (1998). Overreaching and overtraining in endurance athletes. In R.B. Kreider, A.C. Fry, & M.L. O'Toole (Eds.), *Overtraining in sport* (pp. 3–18). Champaign, IL: Human Kinetics.

Pierce, E.F., Jr. (2002). Relationship between training volume and mood states in competitive swimmers during a 24–week season. *Perceptual and Motor Skills, 94,* 1009–1012.

Raglin, J.S. (1993). Overtraining and staleness: Psychometric monitoring of endurance athletes. In R.N. Singer, M. Murphey, & L.K. Tennant (Eds.), *Handbook of research on sport psychology* (pp. 840–850). New York: Macmillan.

Raglin, J.S., Sawamura, S., Alexiou, S., Hassmén, P., & Kenttä, G. (2000). Training practices in 13–18–year–old swimmers: A cross–cultural study. *Pediatric Exercise Science, 12,* 61–70.

Rowbottom, D.G., Keast, D., Garciawebb, P., & Morton, A.R. (1997). Training adaptation and biological changes among well–trained male triathletes. *Medicine & Science in Sports & Exercise, 29,* 1233–1239.

Rowbottom, D.G., Keast, D., & Morton, A.R. (1998). Monitoring and preventing of overreaching and overtraining in endurance athletes. In R.B. Kreider, A.C. Fry, & M.L. O'Toole (Eds.), *Overtraining in sport* (pp. 47–66). Champaign, IL: Human Kinetics.

Shephard, R.J., & Sheck, P.N. (1994). Potential impact of physical activity and sport on the immune system: A brief review. *British Journal of Sports Medicine, 28,* 347–355.

Snyder, A.C. (1998). Overtraining and the glycogen depletion hypothesis. *Medicine & Science in Sports & Exercise, 30,* 1146–1150.

Steinacker, J.M., & Lehmann, M. (2002). Clinical findings and mechanisms of stress and recovery in athletes. In M. Kellmann (Ed.), *Enhancing recovery: Preventing underperformance in athletes* (pp. 103–118). Champaign, IL: Human Kinetics.

Urhausen, A., Gabriel, H.H.W., & Kindermann, W. (1998). Impaired pituitary hormonal response to exhaustive exercise in overtrained endurance athletes. *Medicine & Science in Sports & Exercise, 30,* 407–414.

Urhausen, A., Gabriel, H.H.W., Weiler, B., & Kindermann, W. (1998). Ergometric and psychological findings during overtraining: A long–term follow–up study in endurance athletes. *International Journal of Sports Medicine, 19,* 114–120.

Urhausen, A., & Kindermann, W. (2002). Diagnosis of overtraining: What tools do we have? *Sports Medicine, 32,* 95–102.

Uusitalo, A.L. (2001). Overtraining: Making a difficult diagnosis and implementing targeted treatment. *Physician and Sportsmedicine, 29*(5), 35–50.

Uusitalo, A.L.T., Huttunen, P., Hanin, Y., Uusitalo, A.J., & Rusko, H.K. (1998). Hormonal responses to endurance training and overtraining in female athletes. *Clinical Journal of Sport Medicine, 8,* 178–186.

van Borselen, F., Vos, N.H., & Fry, A.C. (1992). The role of anaerobic exercise in overtraining. *National Strength and Conditioning Association Journal, 14,* 74–79.

Verde, T., Thomas, S., & Shephard, R.J. (1992). Potential markers of heavy training in highly trained distance runners. *British Journal of Sports Medicine, 26,* 167–175.

Wall, S.P., Mattacola, C.B.S., & Levenstein, S. (2003). Sleep efficiency and overreaching in swimmers. *Journal of Sport Rehabilitation, 12,* 1–12.

Weidner, T.G. (1994). Literature review: Upper respiratory illness and sport and exercise. *International Journal of Sports Medicine, 15,* 1–9.

Weinstein, L. (1973). Poliomyelitis: A persistent problem. *New England Journal of Medicine, 288,* 370–371.

Chapter 3

Armstrong, L.E., & VanHeest, J.L. (2002). The unknown mechanism of the overtraining syndrome: Clues from depression and psychoneuroimmunology. *Sports Medicine, 32,* 185–209.

Beil, L. (1988). Of joints and juveniles. *Science News, 134*(12), 190–191.

Botterill, C., & Wilson, C. (2002). Overtraining: Emotional and interdisciplinary dimensions. In M. Kellmann (Ed.), *Enhancing recovery: Preventing underperformance in athletes* (pp. 143–160). Champaign, IL: Human Kinetics.

Brown, R.L., Frederick, E.C., Falsetti, H.L., Burke, E.R., & Ryan, A.J. (1983). Overtraining of athletes: A round table. *Physician and Sportsmedicine, 11*(6), 92–110.

Brustad, R.J., & Ritter–Taylor, M. (1997). Applying social psychological perspectives to the sport psychology consulting process. *The Sport Psychologist, 11*, 107–119.

Budgett, R. (1998). Fatigue and underperformance in athletes: The overtraining syndrome. *British Journal of Sports Medicine, 32*, 107–110.

Committee on Sports Medicine and Fitness (2000, July). Intensive training and sports specialization in young athletes. *Pediatrics, 106*(1), 154–157.

Derman, W., Schwellnus, M.P., Lambert, M.I., Emms, M., Sinclair–Smith, C., Kirby, P., et al. (1997). The 'worn-out athlete': A clinical approach to chronic fatigue in athletes. *Journal of Sports Sciences, 15*(3), 341–351.

Flynn, M.G. (1998). Future research needs and directions. In R.B. Kreider, A.C. Fry, & M.L. O'Toole (Eds.), *Overtraining in sport* (pp. 373–383). Champaign, IL: Human Kinetics.

Foster, C. (1998). Monitoring training in athletes with reference to overtraining syndrome. *Medicine & Science in Sports & Exercise, 30*, 1164–1168.

Fry, R.W., Morton, A.R., & Keast, D. (1991). Overtraining in athletes: An update. *Sports Medicine, 12*, 32–65.

Gould, D., & Dieffenbach, K. (2002). Overt training, underrecovery, and burnout in sport. In M. Kellmann (Ed.), *Enhancing recovery: Preventing underperformance in athletes* (pp. 25–35). Champaign, IL: Human Kinetics.

Gould, D., Guinan, D., Greenleaf, C., & Chung, Y. (2002). A survey of U.S. Olympic coaches: Variables perceived to have influenced athlete performances and coach effectiveness. *The Sport Psychologist, 16*, 229–250.

Gould, D., Guinan, D., Greenleaf, C., Medbery, R., & Peterson, K. (1999). Factors affecting Olympic performance: Perceptions of athletes and coaches from more and less successful teams. *The Sport Psychologist, 13*, 371–394.

Gould, D., Jackson, S.A., & Finch, L.M. (1993). Sources of stress in national champion figure skaters. *Journal of Sport & Exercise Psychology, 15*, 134–159.

Gould, D., Tuffey, S., Udry, E., & Loehr, J. (1997). Burnout in competitive junior tennis players: III. Individual differences in the burnout experience. *The Sport Psychologist, 11*, 257–276.

Hanin, Y.L. (2002). Individual optimal recovery in sport: An application of the IZOF model. In M. Kellmann (Ed.), *Enhancing recovery: Preventing underperformance in athletes* (pp. 199–218). Champaign, IL: Human Kinetics.

Hawley, C.J., & Schoene, R. (2003). Overtraining syndrome. *Physician and Sportsmedicine, 31*(6), 25–31.

Hollander, D.B., & Meyers, M.C. (1995). Psychological factors associated with overtraining: Implications for youth sport coaches. *Journal of Sport Behavior, 18*, 3–20.

Hooper, S.L., & Mackinnon, L.T. (1995). Monitoring overtraining in athletes. *Sports Medicine, 20*, 321–327.

Hooper, S.L., Mackinnon, L.T., Howard, A., Gordon, R.D., & Bachmann, A.W. (1995). Markers for monitoring overtraining and recovery. *Medicine & Science in Sports & Exercise, 27*, 106–112.

Kellmann, M. (2002). Underrecovery and overtraining: Different concepts—similar impact? In M. Kellmann (Ed.), *Enhancing recovery: Preventing underperformance in athletes* (pp. 3–24). Champaign, IL: Human Kinetics.

Kenttä, G., & Hassmén, P. (2002). Underrecovery and overtraining: A conceptual model. In M. Kellmann (Ed.), *Enhancing recovery: Preventing underperformance in athletes* (pp. 57–80). Champaign, IL: Human Kinetics.

Krane, V., Greenleaf, C.A., & Snow, J. (1997). Reaching for gold and the price of glory: A motivational case study of an elite gymnast. *The Sport Psychologist, 11*, 53–71.

Kuipers, H. (1996). How much is too much? Performance aspects of overtraining. *Research Quarterly for Exercise and Sport, 67*(Suppl. 3), S65–S69.

Kuipers, H., & Keizer, H.A. (1988). Overtraining in elite athletes: Review and directions for the future. *Sports Medicine, 6*, 79–92.

Lehmann, M., Foster, C., & Keul, J. (1993). Overtraining in endurance athletes: A brief review. *Medicine & Science in Sports & Exercise, 25*, 854–862.

Levin, S. (1991). Overtraining causes Olympic–sized problems. *Physician and Sportsmedicine, 19*(5), 112–118.

McCann, S. (1995). Overtraining and burnout. In S.M. Murphy (Ed.), *Sport psychology interventions* (pp. 347–369). Champaign, IL: Human Kinetics.

Noh, Y.E., Morris, T., & Andersen, M.B. (2005). Psychosocial factors and ballet injuries. *International Journal of Sport & Exercise Psychology, 1*, 79–90.

O'Toole, M.L. (1998). Overreaching and overtraining in endurance athletes. In R.B. Kreider, A.C. Fry, & M.L. O'Toole (Eds.), *Overtraining in sport* (pp. 3–18). Champaign, IL: Human Kinetics.

Raglin, J.S. (1993). Overtraining and staleness: Psychometric monitoring of endurance athletes. In R.N. Singer, M. Murphey, & L.K. Tennant (Eds.), *Handbook of research on sport psychology* (pp. 840–850). New York: Macmillan.

Raglin, J.S., & Morgan, W.P. (1989). Development of a scale to measure training induced distress. *Medicine and Science in Sport and Exercise, 21*(Suppl.), 60.

Raglin, J.S., & Morgan, W.P. (1994). Development of a scale for use in monitoring training–induced distress in athletes. *International Journal of Sports Medicine, 15*, 84–88.

Scanlan, T.K., Stein, G.L., & Ravizza, K. (1991). An in–depth study of former elite figure skaters: III. Sources of stress. *Journal of Sport & Exercise Psychology, 13*, 103–120.

Uusitalo, A.L. (2001). Overtraining: Making a difficult diagnosis and implementing targeted treatment. *Physician and Sportsmedicine, 29*(5), 35–50.

Wallace, M.A. (1998, July–August). Endurance performance in the heat: Overtraining summit. *Sportscience News.* Retrieved March 20, 2001, from http://sportsci.org/news/news9807/acsmmaw.html

Williams, J.M., & Andersen, M.B. (1998). Psychosocial antecedents of sport injury: Review and critique of the stress and injury model. *Journal of Applied Sport Psychology, 10,* 5–26.

Wrisberg, C.A., & Johnson, M.S. (2002). Quality of life. In M. Kellmann (Ed.), *Enhancing recovery: Preventing underperformance in athletes* (pp. 253–268). Champaign, IL: Human Kinetics.

Chapter 5

Sparkes, A.C. (2002). *Telling tales in sport and physical activity: A qualitative journey.* Champaign, IL: Human Kinetics.

Chapter 7

Kibler, B.W., & Chandler, T.J. (1998). Musculoskeletal and orthopedic considerations. In R.B. Kreider, A.C. Fry, & M.L. O'Toole (Eds.), *Overtraining in sport* (pp. 169–190). Champaign, IL: Human Kinetics.

Kuipers, H. (1996). How much is too much? Performance aspects of overtraining. *Research Quarterly for Exercise and Sport, 67*(Suppl. 3), S65–S69.

Raglin, J.S. (1993). Overtraining and staleness: Psychometric monitoring of endurance athletes. In R.N. Singer, M. Murphey, & L.K. Tennant (Eds.), *Handbook of research on sport psychology* (pp. 840–850). New York: Macmillan.

Chapter 10

Cohen, G.L. (Ed.). (2001). *Women in sport: Issues and controversies* (2nd rev. ed.). Reston, VA: AAHPERD.

Gould, D., Jackson, S.A., & Finch, L.M. (1993). Life at the top: The experience of U.S. national champion figure skaters. *The Sport Psychologist, 7,* 354–374.

Hooper, S.L., Mackinnon, L.T., Howard, A., Gordon, R.D., & Bachmann, A.W. (1995). Markers for monitoring overtraining and recovery. *Medicine & Science in Sports & Exercise, 27,* 106–112.

Kellmann, M. (2002). Underrecovery and overtraining: Different concepts—similar impact? In M. Kellmann (Ed.), *Enhancing recovery: Preventing underperformance in athletes* (pp. 3–24). Champaign, IL: Human Kinetics.

Kibler, B.W., & Chandler, T.J. (1998). Musculoskeletal and orthopedic considerations. In R.B. Kreider, A.C. Fry, & M.L. O'Toole (Eds.), *Overtraining in sport* (pp. 169–190). Champaign, IL: Human Kinetics.

Krane, V., Greenleaf, C.A., & Snow, J. (1997). Reaching for gold and the price of glory: A motivational case study of an elite gymnast. *The Sport Psychologist, 11,* 53–71.

Mackinnon, L.T. (1998). Effects of overreaching and overtraining on immune function. In R.B. Kreider, A.C. Fry, & M.L. O'Toole (Eds.), *Overtraining in sport* (pp. 219–241). Champaign, IL: Human Kinetics.

Messner, M.A. (Ed.). (1990). *Sport, men, and the gender order: Critical feminist perspectives.* Champaign, IL: Human Kinetics.

Morgan, W.P., O'Connor, P.J., Sparling, P.B., & Pate, R.R. (1987). Psychological characterization of the elite female distance runner. *International Journal of Sports Medicine, 8,* 124–131.

Nieman, D.C. (1998). Effects of athletic endurance training on infection rates and immunity. In R.B. Kreider, A.C. Fry, & M.L. O'Toole (Eds.), *Overtraining in sport* (pp. 193–217). Champaign, IL: Human Kinetics.

Rowbottom, D.G., Keast, D., Garciawebb, P., & Morton, A.R. (1997). Training adaptation and biological changes among well–trained male triathletes. *Medicine & Science in Sports & Exercise, 29,* 1233–1239.

Scanlan, T.K., Stein, G.L., & Ravizza, K. (1991). An in–depth study of former elite figure skaters: III. Sources of stress. *Journal of Sport & Exercise Psychology, 13,* 103–120.

Steinacker, J.M., & Lehmann, M. (2002). Clinical findings and mechanisms of stress and recovery in athletes. In M. Kellmann (Ed.), *Enhancing recovery: Preventing underperformance in athletes* (pp. 103–118). Champaign, IL: Human Kinetics.

Uusitalo, A.L.T., Huttunen, P., Hanin, Y., Uusitalo, A.J., & Rusko, H.K. (1998). Hormonal responses to endurance training and overtraining in female athletes. *Clinical Journal of Sport Medicine, 8,* 178–186.

Wrisberg, C.A., & Johnson, M.S. (2002). Quality of life. In M. Kellmann (Ed.), *Enhancing recovery: Preventing underperformance in athletes* (pp. 253–268). Champaign, IL: Human Kinetics.

Young, K., & White, P. (2000). Researching sport injury: Reconstructing dangerous masculinities. In J. McKay, M.A. Messner, & D. Sabo (Eds.), *Masculinities, gender relations, and sport* (pp. 108–126). Thousand Oaks, CA: Sage.

Chapter 11

Berglund, B., & Säfström, H. (1994). Psychological monitoring and modulation of training load of world–class canoeists. *Medicine & Science in Sports & Exercise, 26,* 1036–1040.

Budgett, R. (1998). Fatigue and underperformance in athletes: The overtraining syndrome. *British Journal of Sports Medicine, 32,* 107–110.

Fry, R.W., Grove, J.R., Morton, A.R., Zeroni, P.M., Gaudieri, S., & Keast, D. (1994). Psychological and immunological correlates of acute overtraining. *British Journal of Sports Medicine, 28,* 241–246.

Hooper, S.L., & Mackinnon, L.T. (1995). Monitoring overtraining in athletes. *Sports Medicine, 20,* 321–327.

Hooper, S.L., Mackinnon, L.T., & Hanrahan, S. (1997). Mood states as an indication of staleness and recovery. *International Journal of Sport Psychology, 28,* 1–12.

Hooper, S.L., Mackinnon, L.T., Howard, A., Gordon, R.D., & Bachmann, A.W. (1995). Markers for monitoring overtraining and recovery. *Medicine & Science in Sports & Exercise, 27,* 106–112.

Kellmann, M. (2002). Underrecovery and overtraining: Different concepts—similar impact? In M. Kellmann (Ed.), *Enhancing recovery: Preventing underperformance in athletes* (pp. 3–24). Champaign, IL: Human Kinetics.

Kenttä, G., & Hassmén, P. (1998). Overtraining and recovery: A conceptual model. *Sports Medicine, 26,* 1–16.

Kenttä, G., & Hassmén, P. (2002). Underrecovery and overtraining: A conceptual model. In M. Kellmann (Ed.), *Enhancing recovery: Preventing underperformance in athletes* (pp. 57–80). Champaign, IL: Human Kinetics.

Lehmann, M., Baur, S., Netzer, N., & Gastmann, U. (1997). Monitoring high-intensity endurance training using neuromuscular excitability to recognize overtraining. *European Journal of Applied Physiology & Occupational Physiology, 76,* 187–191.

Lehmann, M., Foster, C., & Keul, J. (1993). Overtraining in endurance athletes: A brief review. *Medicine & Science in Sports & Exercise, 25,* 854–862.

Meyers, A.W., & Whelan, J.P. (1998). A systemic model for understanding psychological influences in overtraining. In R.B. Kreider, A.C. Fry, & M.L. O'Toole (Eds.), *Overtraining in sport* (pp. 335–372). Champaign, IL: Human Kinetics.

Morgan, W.P., Costill, D.L., Flynn, M.G., Raglin, J.S., & O'Connor, P.J. (1988). Mood disturbances following increased training in swimmers. *Medicine & Science in Sports & Exercise, 20,* 408–414.

O'Toole, M.L. (1998). Overreaching and overtraining in endurance athletes. In R.B. Kreider, A.C. Fry, & M.L. O'Toole (Eds.), *Overtraining in sport* (pp. 3–18). Champaign, IL: Human Kinetics.

Steinacker, J.M., & Lehmann, M. (2002). Clinical findings and mechanisms of stress and recovery in athletes. In M. Kellmann (Ed.), *Enhancing recovery: Preventing underperformance in athletes* (pp. 103–118). Champaign, IL: Human Kinetics.

Uusitalo, A.L.T., Huttunen, P., Hanin, Y., Uusitalo, A.J., & Rusko, H.K. (1998). Hormonal responses to endurance training and overtraining in female athletes. *Clinical Journal of Sport Medicine, 8,* 178–186.

Chapter 12

Martin, D.T., Andersen, M.B., & Gates, W. (2000). Using Profile of Mood States (POMS) to monitor high-intensity training in cyclists: Groups versus case studies. *The Sport Psychologist, 14,* 138–156.

Morgan, W.P., O'Connor, P.J., Ellickson, K.A., & Bradley, P.W. (1988). Personality structure, mood states, and performance in elite male distance runners. *International Journal of Sport Psychology, 19,* 247–263.

Morgan, W.P., O'Connor, P.J., Sparling, P.B., & Pate, R.R. (1987). Psychological characterization of the elite female distance runner. *International Journal of Sports Medicine, 8,* 124–131.

Index

Note: The italicized *f* and *t* following page numbers refer to figures and tables, respectively.

About the Authors

Sean O. Richardson, PhD, completed his doctoral work in sport psychology at Victoria University (Melbourne, Australia) in 2006. His dissertation research focused on the risk factors for athletic overtraining, stress–life imbalance, and injury.

Richardson has also been a competitive athlete most of his life. He has pursued windsurfing and rowing at national and international levels, along with several other sports at the state and provincial levels, including road and track cycling, downhill skiing, and volleyball. He has had personal experiences with injury related to overtraining behaviors, missing out on two chances to make the Canadian Olympic team in rowing because of injury.

Throughout Australia and Canada, Richardson now serves as a sport and performance psychologist in the areas of performance enhancement, injury and illness prevention, rehabilitation, and stress–life balance for numerous sport and performing arts groups as well as health care and business professionals. He regularly delivers seminars on optimal recovery and injury prevention to athletes, coaches, performing artists, and teachers of all levels, from novice to professional.

Mark B. Andersen, PhD, is a professor in the School of Human Movement, Recreation and Performance at Victoria University (Melbourne, Australia). He received his PhD in psychology with a minor in exercise and sport sciences from the University of Arizona at Tucson in 1988.

In 1994 Andersen received the Dorothy V. Harris Memorial Award for excellence as a young scholar and practitioner in applied sport psychology from the Association for Applied Sport Psychology. He has published more than 50 articles in refereed journals and more than 65 book chapters and proceedings. He has edited two other Human Kinetics books: *Doing Sport Psychology* and *Sport Psychology in Practice*. He is also the editor of the text *Psychology in the Physical and Manual Therapies* published by Churchill/Livingstone. Andersen is a member of the International Society of Sport Psychology and a charter member of the Association for Applied Sport Psychology.

Tony Morris, PhD, is a professor in the School of Human Movement, Recreation and Performance at Victoria University (Melbourne, Australia). He received his doctoral degree from the University of Leeds in England in 1984.

Morris has published more than 30 books, monographs, and book chapters and more than 80 articles in refereed journals. He presents his research worldwide, having been invited to speak at conferences in the United Kingdom, Greece, Australia, and throughout Southeast Asia. Morris is a graduate member of the British Psychological Society, a full member of the Australian Psychological Society, and a founding member of the Board of Sport Psychologists in the Australian Psychological Society. He is also a member of the British Society of Sport Psychology, British Association of Sport and Exercise Sciences, North American Society for the Psychology of Sport and Physical Activity, International Society of Sport Psychology, Association for Applied Sport Psychology, British Society of Experimental and Clinical Hypnosis, and the Sport Psychology Association of Australia and New Zealand.

He has served on the editorial board for a number of journals, including the *International Journal of Sport Psychology*, *Journal of Sports Sciences*, *International Journal of Sport and Exercise Psychology*, and *Research in Sports Medicine: An International Journal*. Morris is also the associate editor for *Australian Psychologist*.

You'll find
other outstanding
educational resources at

www.HumanKinetics.com

In the U.S. call

1-800-747-4457

Australia...08 8372 0999
Canada 1-800-465-7301
Europe...+44 (0) 113 255 5665
New Zealand.......................................0064 9 448 1207

HUMAN KINETICS
The Premier Publisher for Sports & Fitness
P.O. Box 5076 • Champaign, IL 61825-5076 USA